Summable Spaces and Their Duals, Matrix Transformations and Geometric Properties

Monographs and Research Notes in Mathematics

About the Series

This series is designed to capture new developments and summarize what is known over the entire field of mathematics, both pure and applied. It will include a broad range of monographs and research notes on current and developing topics that will appeal to academics, graduate students, and practitioners. Interdisciplinary books appealing not only to the mathematical community, but also to engineers, physicists, and computer scientists are encouraged.

This series will maintain the highest editorial standards, publishing well-developed monographs as well as research notes on new topics that are final, but not yet refined into a formal monograph. The notes are meant to be a rapid means of publication for current material where the style of exposition reflects a developing topic.

Neutrices and External Numbers
A Flexible Number System
Dinis Bruno, Imme van den Berg

Extending Structures
Fundamentals and Applications
Ana Agore, Gigel Militaru

Inverse Scattering Problems and Their Application to Non-Linear Integrable Equations
Pham Loi Vu

Spectral Methods Using Multivariate Polynomials on the Unit Ball
Kendall Atkinson, David Chien, and Olaf Hansen

Glider Representations
Frederik Caenepeel, Fred Van Oystaeyen

Lattice Point Identities and Shannon-Type Sampling
Willi Freeden, M. Zuhair Nashed

Summable Spaces and Their Duals, Matrix Transformations and Geometric Properties
Feyzi Başar, Hemen Dutta

For more information about this series please visit: https://www.crcpress.com/Chapman--HallCRC-Monographs-and-Research-Notes-in-Mathematics/book-series/CRCMONRESNOT

Summable Spaces and Their Duals, Matrix Transformations and Geometric Properties

Feyzi Başar
İnönü University

Hemen Dutta
Gauhati University

CRC Press
Taylor & Francis Group
Boca Raton London New York

CRC Press is an imprint of the
Taylor & Francis Group, an **informa** business
A CHAPMAN & HALL BOOK

CRC Press
Taylor & Francis Group
6000 Broken Sound Parkway NW, Suite 300
Boca Raton, FL 33487-2742

Printed on acid-free paper

International Standard Book Number-13: 978-0-8153-5177-1 (Hardback)

Library of Congress Control Number: 2019051952

Visit the Taylor & Francis Web site at
http://www.taylorandfrancis.com

and the CRC Press Web site at
http://www.crcpress.com

Dedicated by the authors to their parents and by the first author to his dearest daughter Saliha Zehra in affection and devotion.

F.B. & H.D.

Contents

Preface

This book is intended for graduate students, researchers and educators in mathematics with interest in mathematical analysis in general and in summability theory in particular. It covers several topics on normed and paranormed sequence spaces derived as the domain of triangles, matrix transformations, computation of duals, geometric properties and several other related topics. The book consists of five chapters and is organized as follows:

Chapter 1 presents some basic definitions, notations and various basic ideas to provide the readers with an overview on the fundamentals required to understand the main topics and ideas discussed in the remaining chapters.

Chapter 2 emphasizes the domains of Nörlund means, domains of the Euler-Cesàro difference matrix and domains of the band matrix $\widehat{F}$ defined by the Fibonacci sequence in the classical sequence spaces.

Chapter 3 studies the sequence spaces $\ell_\infty(\widetilde{B},p)$, $c(\widetilde{B},p)$ and $c_0(\widetilde{B},p)$ generated by the domains of double sequential band matrix $B(\widetilde{r},\widetilde{s})$ in the Maddox's spaces $\ell_\infty(p)$, $c(p)$, $c_0(p)$ and $\ell(p)$, and investigated their topological properties.

Chapter 4 introduces the paranormed Nörlund sequence spaces $N^t(p)$ and determines its alpha-, beta- and gamma-duals. We characterize the classes of matrix transformations from the space $N^t(p)$ into any given sequence space μ and from a given sequence space μ into the space $N^t(p)$. Finally, we characterize the rotundity of the space $N^t(p)$ and present some results related to this concept.

Chapter 5 deals with various ways to extend Orlicz sequence spaces and discusses relevant algebraic and topological properties of such spaces.

The presentation of each topic is made to a reasonable extent and readers will find at least a hint to solutions of all problems discussed. At the end of each chapter, several research problems have been mentioned and those should be helpful for early career researchers for further study and designing new research proposals.

The authors would like to thank Dr. Medine Yeşilkayagil, Department of Mathematics, Uşak University, Uşak, Turkey, for her careful reading and valuable suggestions on the earlier version of this book. Those suggestions were very helpful for revising and improving this book, as well as significant guiding factors for further research.

Finally, we are very grateful to the reviewers, several colleagues and friends, and the editorial staff at Taylor & Francis for their cooperation in every step of the publishing process.

Feyzi Başar
İstanbul, Turkey

Hemen Dutta
Guwahati, India
October, 2019

Authors

Dr. Feyzi Başar is a Professor Emeritus since July 2016 at İnönü University, Turkey. He has published an e-book for graduate students and researchers and more than 150 scientific papers in the field of summability theory, sequence spaces, FK-spaces, Schauder bases, dual spaces, matrix transformations, spectrum of certain linear operators represented by a triangle matrix over some sequence space, the alpha-, beta- and gamma-duals and some topological properties of the domains of some double and four-dimensional triangles in certain spaces of single and double sequences, sets of the sequences of fuzzy numbers, and multiplicative calculus. He has guided 17 MA and 10 PhD students and served as a referee for 121 international scientific journals. He is the member of editorial boards of 21 scientific journals. He is also a member of scientific committees of 17 mathematics conferences, delivered talks at 14 different universities as an invited speaker and participated in more than 70 mathematics symposiums with papers.

Dr. Hemen Dutta has been rendering his services as a regular faculty member in mathematics at Gauhati University, India. His research areas include functional analysis, mathematical modelling, and many more. He has to his credit over 100 research papers and book chapters as well as 10 books so far. He has acted as a resource person at different academic events, and delivered invited talks at national and international levels. He has visited several foreign countries by invitation for research collaboration and lectures. He has conducted five academic events and worked at several conferences in different capacities. He is also involved with publishing thematic issues in journals and book series. He is actively involved in popularizing mathematics education in different ways from high school to higher levels. He was honorary joint secretary of the Assam Academy of Mathematics for two years during 2014-2015. He is also a life member of several mathematical societies. He has authored articles for newspaper, scientific magazines, science portals, popular books, and many more.

List of Abbreviations and Symbols

$\overline{A}$	:	the closure of a set A
$\mathbb{N}$	:	set of natural numbers, i.e., $\mathbb{N} = \{0, 1, 2, \ldots\}$
$\mathbb{N}_k$	:	set of integers which are greater than or equal to $k \in \mathbb{N}$
$\mathbb{Z}$	:	set of integers, i.e., $\mathbb{Z} = \{\ldots, -2, -1, 0, 1, 2, \ldots\}$
$\mathbb{Q}$	:	set of rational numbers
$\mathbb{R}$	:	set of real numbers, the real field
$\mathbb{R}^+$	:	set of non-negative real numbers
$\mathbb{C}$	:	set of complex numbers, the complex field
$\mathbb{R}^n$	:	n-dimensional Euclidean space
$\mathbb{C}^n$	:	n-dimensional complex Euclidean space
$[a]$	:	integer part of a number a
$\mathcal{F}$	:	collection of all finite subsets of $\mathbb{N}$
e	:	sequence whose elements are all equal to 1
$e^{(k)}$	:	sequences whose only non-zero term is a 1 in k^{th} place for each $k \in \mathbb{N}$
δ_{ij}	:	Kronecker delta which is $= 1$ if $i = j$ and $= 0$ if $i \neq j$
ω	:	space of all sequences over the complex field
$\mathcal{U}$	:	set of all sequences $u = (u_k)$ such that $u_k \neq 0$ for all $k \in \mathbb{N}$
ϕ	:	set of all finitely non-zero sequences
ℓ_∞	:	space of bounded sequences over the complex field
f	:	space of almost convergent sequences over the complex field
$[f]$	:	space of all strongly almost convergent sequences over the complex field
f_0	:	space of almost null sequences over the complex field
c	:	space of convergent sequences over the complex field
c_0	:	space of null sequences over the complex field
ℓ_M	:	Orlicz sequence space
ℓ_1	:	space of absolutely summable sequences over the complex field
ℓ_p	:	space of absolutely p-summable sequences over the complex field
fs	:	space of almost convergent series over the complex field
bs	:	space of bounded series over the complex field
fs_0	:	space of series almost converging to zero over the complex field
cs	:	space of convergent series over the complex field
cs_0	:	space of series converging to zero over the complex field
bv	:	space of sequences of bounded variation over the complex field

bv_0	: space of sequences of both bounded variation and null over the complex field
bv_p	: space of sequences of p-bounded variation over the complex field
$\ell_\infty(p)$	: space of all sequences (x_k) such that $\sup_{k\in\mathbb{N}} \|x_k\|^{p_k} < \infty$
$c(p)$	: space of all sequences (x_k) such that $\|x_k - l\|^{p_k} \to 0$, as $k \to \infty$
$c_0(p)$	: space of all sequences (x_k) such that $\|x_k\|^{p_k} \to 0$, as $k \to \infty$
$\ell(p)$	: space of all sequences (x_k) such that $\sum_k \|x_k\|^{p_k} < \infty$
X^α	: alpha dual of a sequence space X
X^β	: beta dual of a sequence space X
X^γ	: gamma dual of a sequence space X
X^*	: continuous dual of a sequence space X
X^f	: f-dual of a sequence space X
$x^{[n]}$	: n^{th} section of a sequence $x = (x_k)$
$\{(Ax)_n\}$	: A-transform of a sequence x
A_n	: the sequence in the nth row of an infinite matrix $A = (a_{nk})$
c_A	: convergence domain of a matrix A
$\chi(A)$	: characteristic of a matrix A
M	: Orlicz function
L	: Banach limit
S	: shift operator
$S = (s_{nk})$	: summation matrix
Δ	: forward difference operator, i.e., $(\Delta x)_k = x_k - x_{k+1}$ and $(\Delta^2 x)_k = \Delta(x_k - x_{k+1})$
∇	: backward difference operator, i.e., $(\nabla x)_k = x_k - x_{k-1}$
g_1	: paranorm on the spaces $\ell_\infty(p)$, $c(p)$ and $c_0(p)$
g_2	: paranorm on the space $\ell(p)$
g_3	: paranorm on the spaces $\omega_\infty(p)$, $\omega(p)$ and $\omega_0(p)$
g_4	: paranorm on the spaces $f_0(p)$ and $f(p)$
g_5	: paranorm on the spaces $bs(p)$, $cs(p)$ and $cs_0(p)$
g_6	: paranorm on the spaces $\ell_\infty(\widetilde{B}, p)$, $c(\widetilde{B}, p)$ and $c_0(\widetilde{B}, p)$
g_7	: paranorm on the space $N^t(p)$
h	: paranorm on the space $\ell(\widetilde{B}, p)$
C	: Cesàro matrix of order one
E_1	: original Euler matrix
E_q	: Euler mean of order q
E^r	: Euler-Knopp matrix of order r
N^t	: Nörlund mean with respect to the sequence $t = (t_k)$
R^q	: Riesz mean with respect to the sequence $q = (q_k)$
θ	: zero vector in a linear space X
$(\lambda : \mu)$	: class of all matrices A such that $A : \lambda \to \mu$
$(c : c)$	: class of conservative matrices
$(c : c; p)$	: class of Teoplitz matrices
$(c : c)_{reg}$	: class of regular matrices
$(cs : c; p)$	: class of series to sequence regular matrices
$(c : v_\sigma)$	: class of sequence to sequence sigma-conservative matrices

$(c : v_\sigma)_{reg}$	:	class of sequence to sequence sigma-regular matrices
$(c : f)$	:	class of almost conservative matrices
$(c : f)_{reg}$	:	class of almost regular matrices
$(f : c)$	:	class of strongly conservative matrices
$(f : c; p)$	:	class of strongly regular matrices
$(\ell_\infty : c)$	:	class of Schur (coercive) matrices
$(\ell_\infty : f)$	:	class of sequence to sequence almost coercive matrices
$(\ell_\infty : fs)$	:	class of sequence to series almost coercive matrices
$(bs : f)$	:	class of series to sequence almost coercive matrices
$(bs : fs)$	:	class of series to series almost coercive matrices
$\emptyset$	:	empty set
$(AB)_{ij}$	:	i^{th} row and j^{th} column entry of the matrix product AB
I	:	unit matrix

Chapter 1

Linear Sequence Spaces and Matrix Domains in Sequence Spaces

Abstract. *In this chapter, we present some basic definitions and notations used throughout the book necessary for understanding the main topics and ideas discussed in the remaining chapters.*

Keywords. *Linear sequence spaces, Metric sequence spaces, Classical sequence spaces, Normed linear spaces, Jensen's inequality, K-space, AK-space, BK-space, FK-space, Schauder basis, Separability, Reflexivity.*

1.1 Linear Sequence Spaces

By ω, we mean the set of all sequences with complex terms, i.e.,

$$\omega := \{x = (x_k) : x_k \in \mathbb{C} \text{ for all } k \in \mathbb{N}\},$$

where $\mathbb{N}$ denotes the set of natural numbers, i.e., $\mathbb{N} = \{0, 1, 2, \ldots\}$. It is a routine verification that ω is a linear space with respect to the coordinatewise addition and scalar multiplication of sequences which are defined, as usual, by

$$x + y = (x_k) + (y_k) = (x_k + y_k) \text{ and } \alpha x = \alpha(x_k) = (\alpha x_k),$$

respectively, where $x = (x_k)$, $y = (y_k) \in \omega$ and $\alpha \in \mathbb{C}$, the set of complex numbers. By a *sequence space*, we understand a linear subspace of the space ω.

Let $e = (e_k)$ and $e^{(n)} = (e_k^{(n)})$ for each $n \in \mathbb{N}$ be the sequences with $e_k = 1$ for all $k \in \mathbb{N}$, and $e_k^{(n)} = \delta_{nk}$ for all $k, n \in \mathbb{N}$, where δ_{nk} denotes the Kronecker delta defined by

$$\delta_{nk} = \begin{cases} 1 & , \quad n = k, \\ 0 & , \quad n \neq k \end{cases}$$

for all $n, k \in \mathbb{N}$.

We assume throughout unless stated otherwise that $1 \leq p < \infty$ and q is the conjugate number of p, that is, $q = \infty$ if $p = 1$ and $q = p/(p-1)$ for $1 < p < \infty$, and use the convention that any term with a negative subscript is equal to naught.

1.1.1 Metric Sequence Spaces

Prior to giving examples of metric sequence spaces, we note the definition of the concepts of a metric and a metric space.

Definition 1.1.1. *Let X be a non-empty set and d be a distance function from $X \times X$ to the set $\mathbb{R}^+$ of non-negative real numbers. Then the pair (X, d) is called a metric space and d is a metric for X, if the following metric axioms are satisfied for all elements $x, y, z \in X$:*

(M.1) $d(x, y) = 0$ *if and only if* $x = y$.

(M.2) $d(x, y) = d(y, x)$, *(the symmetry property).*

(M.3) $d(x, z) \leq d(x, y) + d(y, z)$, *(the triangle inequality).*

Now, we can present some examples of metric sequence spaces:

1.1.2 The Space ω

The most popular metric d_ω on the space ω is defined by

$$d_\omega(x, y) = \sum_k \frac{|x_k - y_k|}{2^k(1 + |x_k - y_k|)}; \quad x = (x_k), \; y = (y_k) \in \omega.$$

For simplicity in notation, here and in what follows, the summation without limits runs from 0 to ∞. It is obvious that another metric for the space ω can be derived from d_ω replacing 2^{-k} by μ_k such that $\sum_k \mu_k$ is a convergent series with positive terms.

1.1.3 The Space ℓ_∞

By ℓ_∞, we denote the space of bounded sequences, that is,

$$\ell_\infty := \left\{ x = (x_k) \in \omega : \sup_{k \in \mathbb{N}} |x_k| < \infty \right\}.$$

The natural metric d_∞ on the space ℓ_∞ is defined by

$$d_\infty(x, y) = \sup_{k \in \mathbb{N}} |x_k - y_k|; \quad x = (x_k), \; y = (y_k) \in \ell_\infty.$$

1.1.4 The Spaces f and f_0

The concept of *almost convergence* arises as a generalization of that of ordinary convergence. Banach [19] proved the existence of a functional L on ℓ_∞ satisfying the following conditions for all $x, y \in \ell_\infty$ and all scalars λ and μ:

(i) $L(\lambda x + \mu y) = \lambda L(x) + \mu L(y)$,

(ii) $x_k \geq 0$ for all $k \in \mathbb{N}$ implies $L((x_k)_{k=0}^{\infty}) \geq 0$,

(iii) $L((x_{k+n})_{k=0}^{\infty}) = L((x_k)_{k=0}^{\infty})$ for all $n \in \mathbb{N}$,

(iv) $L(e) = 1$.

Lorentz [81] defined a *Banach limit* to be any functional on ℓ_∞ satisfying the condition in (i)–(iv), and a sequence $x = (x_k) \in \ell_\infty$ is said to be almost convergent to the generalized limit α if all Banach limits of x coincide and are equal to α, [81]. This is denoted by $f - \lim x_k = \alpha$. The *shift operator* P is defined on ω by $P_n(x) = x_{n+1}$ for all $n \in \mathbb{N}$. Let P^i be the composition of P with itself i times and write for a sequence $x = (x_k)$

$$t_{mn}(x) = \frac{1}{m+1}\sum_{i=0}^{m} P_n^i(x) \quad \text{for all} \quad m, n \in \mathbb{N}.$$

Lorentz [81] proved that $f - \lim x_k = \alpha$ if and only if $\lim_{m\to\infty} t_{mn}(x) = \alpha$, uniformly in n. It is well known that a convergent sequence is almost convergent such that its ordinary and generalized limits are equal. Therefore, one can define the spaces f and f_0 of almost convergent and almost null sequences, as follows

$$f := \left\{(x_k) \in \ell_\infty : \lim_{m\to\infty} \sum_{k=0}^{m} \frac{x_{n+k}}{m+1} = \alpha \text{ uniformly in } n \text{ for some } \alpha \in \mathbb{C}\right\},$$

$$f_0 := \left\{(x_k) \in \ell_\infty : \lim_{m\to\infty} \sum_{k=0}^{m} \frac{x_{n+k}}{m+1} = 0 \text{ uniformly in } n\right\}.$$

The metric d_∞ is also a metric for the spaces f and f_0. For more detail on the Banach limit, the reader may refer to Çolak and Çakar [44], and Das [46].

1.1.5 The Spaces c and c_0

By c and c_0, we denote the spaces of convergent and null sequences, i.e.,

$$c := \left\{x = (x_k) \in \omega : \lim_{k\to\infty} |x_k - l| = 0 \text{ for some } l \in \mathbb{C}\right\},$$

$$c_0 := \left\{x = (x_k) \in \omega : \lim_{k\to\infty} x_k = 0\right\}.$$

The metric d_∞ is also a metric for the spaces c and c_0. It is trivial that since the concepts supremum and maximum are equivalent on the space c_0, the metric d_∞ is reduced to the metric d_0 defined by

$$d_0(x, y) = \max_{k\in\mathbb{N}} |x_k - y_k|; \quad x = (x_k),\ y = (y_k) \in c_0$$

on the space c_0.

1.1.6 The Space ℓ_p

By ℓ_p, we denote the space of absolutely p-summable sequences, that is,

$$\ell_p := \left\{x = (x_k) \in \omega : \sum_k |x_k|^p < \infty\right\}, \quad (0 < p < \infty).$$

In the case $1 \leq p < \infty$, the metric d_p on the space ℓ_p is given by

$$d_p(x, y) = \left(\sum_k |x_k - y_k|^p\right)^{1/p}; \quad x = (x_k),\ y = (y_k) \in \ell_p.$$

Also in the case $0 \leq p < 1$, the metric $\widetilde{d_p}$ on the space ℓ_p is given by

$$\widetilde{d_p}(x, y) = \sum_k |x_k - y_k|^p; \quad x = (x_k),\ y = (y_k) \in \ell_p.$$

1.1.7 The Space bs

By bs, we mean the space of bounded series, that is,

$$bs := \left\{x = (x_k) \in \omega : \sup_{n \in \mathbb{N}} \left|\sum_{k=0}^{n} x_k\right| < \infty\right\}.$$

The natural metric on the space bs is defined by

$$d(x, y) = \sup_{n \in \mathbb{N}} \left|\sum_{k=0}^{n} (x_k - y_k)\right|; \quad x = (x_k),\ y = (y_k) \in bs. \tag{1.1}$$

1.1.8 The Spaces cs and cs_0

By cs and cs_0, we denote the spaces of convergent series and the space of the series converging to zero, i.e.,

$$\begin{aligned} cs &:= \left\{x = (x_k) \in \omega : \lim_{n \to \infty} \left|\sum_{k=0}^{n} x_k - l\right| = 0 \text{ for some } l \in \mathbb{C}\right\}, \\ cs_0 &:= \left\{x = (x_k) \in \omega : \lim_{n \to \infty} \left|\sum_{k=0}^{n} x_k\right| = 0\right\}. \end{aligned}$$

The relation d defined by (1.1) is the natural metric on the spaces cs and cs_0.

1.1.9 The Spaces bv and bv_1

The spaces bv and bv_1 of sequences of bounded variation are defined by

$$\begin{aligned} bv &:= \left\{x = (x_k) \in \omega : \sum_k |x_k - x_{k+1}| < \infty\right\}, \\ bv_1 &:= \left\{x = (x_k) \in \omega : \sum_k |x_k - x_{k-1}| < \infty\right\}. \end{aligned}$$

Define the forward and backward difference operators Δ and ∇ by $(\Delta u)_k = u_k - u_{k+1}$ and $(\nabla u)_k = u_k - u_{k-1}$ for all $k \in \mathbb{N}$ with $u_{-1} = 0$. Then, the natural metrics on the spaces bv and bv_1 are defined by

$$d(x,y) = |x_0 - y_0| + \sum_{k=1}^{\infty} |(\Delta(x-y))_k|; \quad x = (x_k),\ y = (y_k) \in bv,$$
$$d(x,y) = \sum_k |(\nabla(x-y))_k|; \quad x = (x_k),\ y = (y_k) \in bv_1.$$

1.1.10 The Spaces ω_0^p, ω^p and ω_∞^p

The spaces ω_0^p, ω^p and ω_∞^p of sequences that are strongly summable to zero, summable and bounded of index $p \geq 1$ by the Cesàro method of order 1 are defined by

$$\begin{aligned} \omega_0^p &:= \left\{x = (x_k) \in \omega : \lim_{n\to\infty} \frac{1}{n+1} \sum_{k=0}^{n} |x_k|^p = 0\right\}, \\ \omega^p &:= \left\{x = (x_k) \in \omega : \lim_{n\to\infty} \frac{1}{n+1} \sum_{k=0}^{n} |x_k - l|^p = 0 \text{ for some } l \in \mathbb{C}\right\}, \\ \omega_\infty^p &:= \left\{x = (x_k) \in \omega : \sup_{n\in\mathbb{N}} \frac{1}{n+1} \sum_{k=0}^{n} |x_k|^p < \infty\right\}. \end{aligned}$$

The metric $\widehat{d}_p$ on the spaces ω_0^p, ω^p and ω_∞^p is defined by

$$\widehat{d}_p(x,y) = \begin{cases} \sup_{n\in\mathbb{N}} \frac{1}{n+1} \sum_{k=0}^{n} |x_k - y_k|^p & , \quad 0 < p < 1, \\ \sup_{n\in\mathbb{N}} \left(\frac{1}{n+1} \sum_{k=0}^{n} |x_k - y_k|^p\right)^{1/p} & , \quad 1 \leq p < \infty. \end{cases}$$

As an easy consequence of Hölder's inequality, the inclusion $\omega^q \subset \omega^p$ holds when $0 < p < q$.

We now give the definition of Fréchet space.

Definition 1.1.2. *Let X be a real or complex linear space and d be a metric on X. Then, (X,d) is said to be a linear metric space, if the algebraic operations* *addition (+) and scalar multiplication (·) on X are continuous. A complete linear metric space is said to be a Fréchet space.*

One can show that the metric space ω of all complex sequences forms a Fréchet space with respect to the metric d_ω.

1.1.11 Normed Sequence Spaces

We begin with giving the definitions of a norm and a p-norm.

Definition 1.1.3. *Let X be a real or complex linear space and $\|\cdot\|$ be a function from X to the set $\mathbb{R}^+$ of non-negative real numbers. Then the pair $(X, \|\cdot\|)$ is called a normed space and $\|\cdot\|$ is a norm for X, if the following norm axioms are satisfied for all elements $x, y \in X$ and for all scalars α:*

(N.1) $\|x\| = 0$ *if and only if* $x = \theta$.

(N.2) $\|\alpha x\| = |\alpha|\|x\|$, *(the absolute homogeneity property).*

(N.3) $\|x + y\| \leq \|x\| + \|y\|$, *(the triangle inequality).*

Definition 1.1.4. *Let X be a real or complex linear space, $\|\cdot\|$ be a function from X to the set $\mathbb{R}^+$ of non-negative real numbers and $p > 0$. Then, the pair $(X, \|\cdot\|)$ is called a p-normed space and $\|\cdot\|$ is a p-norm for X, if the following p-norm axioms are satisfied for all elements $x, y \in X$ and for all scalars α:*

(P.1) $\|x\| = 0$ *if and only if* $x = \theta$.

(P.2) $\|\alpha x\| = |\alpha|^p\|x\|$.

(P.3) $\|x + y\| \leq \|x\| + \|y\|$.

Now, following Malkowsky [91], we can give the examples of normed and p-normed sequence spaces. First, consider the following relations which are derived from the corresponding metrics given in Subsections 1.1.2-1.1.10 by putting, as usual, $\|x\| = d(x, \theta)$:

$$
\begin{aligned}
\|x\|_\infty &= \sup_{k\in\mathbb{N}} |x_k|, \quad \|x\|_0 = \max_{k\in\mathbb{N}} |x_k|, \\
\|x\|_p &= \begin{cases} \sum_k |x_k|^p &, \quad 0 < p < 1, \\ \left(\sum_k |x_k|^p\right)^{1/p} &, \quad 1 \leq p < \infty, \end{cases} \\
\|x\|_{bv} &= |x_0| + \sum_{k=1}^{\infty} |x_k - x_{k+1}|, \quad \|x\|_{bv_1} = \sum_k |x_k - x_{k-1}|, \\
\|x\|_{cs} &= \|x\|_{bs} = \sup_{k\in\mathbb{N}} \left|\sum_{j=0}^{k} x_j\right|, \qquad (1.2) \\
\widehat{\|x\|}_p &= \begin{cases} \sup_{n\in\mathbb{N}} \frac{1}{n+1} \sum_{k=0}^{n} |x_k|^p &, \quad 0 < p < 1, \\ \sup_{n\in\mathbb{N}} \left(\frac{1}{n+1} \sum_{k=0}^{n} |x_k|^p\right)^{1/p} &, \quad 1 \leq p < \infty. \end{cases}
\end{aligned}
$$

It is not hard to show that ℓ_∞ and c are Banach spaces with the norm $\|\cdot\|_\infty$. Additionally c_0 and bv are Banach spaces with the norms $\|\cdot\|_0$ and $\|\cdot\|_{bv}$. The spaces bs and cs are Banach spaces with the norm $\|\cdot\|_{bs}$. In the case $1 \leq p < \infty$, the spaces ℓ_p and ω_0^p, ω^p, ω_∞^p are Banach spaces with the norms $\|\cdot\|_p$ and $\widehat{\|\cdot\|}_p$, respectively. In the case $0 < p < 1$, the spaces ℓ_p and ω_0^p, ω^p, ω_∞^p are complete p-normed spaces with the p-norms $\|\cdot\|_p$ and $\widehat{\|\cdot\|}_p$, respectively.

Consider *Jensen's inequality*

$$\left(\sum_{k=0}^{n} |x_k|^r\right)^{1/r} < \left(\sum_{k=0}^{n} |x_k|^s\right)^{1/s},$$

where $r > s > 0$ and $x_0, x_1, \ldots, x_n \in \mathbb{C}$. Jensen's inequality gives the inclusion $\ell_r \supset \ell_p$ for $r > p$.

A sequence space λ with a linear topology is called a *K-space* provided each of the maps p_i defined by

$$\begin{array}{rccc} p_i : & \lambda & \longrightarrow & \mathbb{C} \\ & x = (x_i) & \longmapsto & p_i(x) = x_i \end{array}$$

is continuous for all $i \in \mathbb{N}$. A K-space λ is called an *FK-space* provided λ is a complete linear metric space. An FK-space whose topology is normable is called a *BK-space*.

Given an FK-space $\lambda \supset \phi$, we denote the nth *section of a sequence* $x = (x_k) \in \lambda$ by $x^{[n]} = \sum_{k=0}^{n} x_k e^{(k)}$, and we say that x has the property

- *AK* if $x^{[n]} \to x$ as $n \to \infty$ (Abschnittskonvergenz),
- *AB* if $\left(x^{[n]}\right)$ is bounded (Abschnittsbeschränktheit),
- *AD* if $x \in \overline{\phi}$ (abschnittsdicht),
- *KB* if the set $\{x_k e^{(k)}\}$ is bounded in λ (koordinatenweise beschränkt).

If one of these properties holds for every $x \in \lambda$ then we say that the space λ has that property (cf. [59]). It is trivial that AK implies AD and AK iff $AB + AD$. For example, c_0 and ℓ_p are AK-spaces, and c and ℓ_∞ are not AD-spaces.

If a normed sequence space λ contains a sequence (b_n) with the property that for every $x \in \lambda$ there is a unique sequence of scalars (α_n) such that

$$\lim_{n\to\infty} \left\| x - (\alpha_0 b_0 + \alpha_1 b_1 + \cdots + \alpha_n b_n) \right\| = 0$$

then (b_n) is called a *Schauder basis* (or briefly *basis*) for λ. The series $\sum_k \alpha_k b_k$ which has the sum x is then called the expansion of x with respect to (b_n), and written as $x = \sum_k \alpha_k b_k$. The concepts of Schauder and algebraic bases coincide for finite dimensional spaces. Nevertheless, there are linear spaces without a Schauder basis.

Example 1.1.5. *The sequence* $\{e^{(k)}\}_{k\in\mathbb{N}}$ *is a Schauder basis for the space* ω *under its natural paranorm* g_ω (cf. [104, Example 23, p. 147]) *defined by*

$$g_\omega(x) = \sum_k \frac{|x_k|}{2^k(1+|x_k|)}, \quad x = (x_k) \in \omega$$

and any $x = (x_k) \in \omega$ *has a unique representation of the form* $\lim_{n\to\infty} x^{[n]} = x$.

Example 1.1.6. [134, Example 6.3.5, p. 88] The sequence

$$\{e, e^{(0)}, e^{(1)}, \dots\}$$

is a Schauder basis for the space c and any $x = (x_k) \in c$ has a unique representation of the form

$$x = \alpha \cdot e + \sum_k (x_k - \alpha)e^{(k)}, \quad \text{where } \alpha = \lim_{k\to\infty} x_k. \tag{1.3}$$

If λ is an FK-space, $\phi \subset \lambda$ and $(e^{(k)})$ is a basis for λ then λ is said to have AK property, where $e^{(k)}$ is a sequence whose only term in k^{th} place is 1 and the others are zero for each $k \in \mathbb{N}$ and $\phi = span\{e^{(k)}\}$. If ϕ is dense in λ, then λ is called AD-space thus AK implies AD.

Now, we give the following definition as it concern with the perfectness, normality and monotonity of a sequence space (see [64, p. 48]):

Definition 1.1.7. *Let* λ *be a sequence space. Then,*

(i) λ *is called perfect if* $\lambda = \lambda^{\alpha\alpha}$, *where* $\lambda^{\alpha\alpha}$ *is the* $\alpha-$*dual of the space* λ^α.

(ii) λ *is called solid or normal if and only if*

$$\widetilde{\lambda} := \{(u_k) \in \omega : \exists (x_k) \in \lambda \text{ such that } |u_k| \le |x_k| \text{ for all } k \in \mathbb{N}\} \subset \lambda.$$

(iii) For a sequence J *of* $\mathbb{N}$ *and a sequence space* λ, *we define* λ_J *by*

$$\lambda_J := \{x = (x_i) : \text{there is a } y = (y_i) \in \lambda \text{ with } x_i = y_{n_i} \text{ for all } n_i \in J\}$$

and call λ_J *the* J*-stepspace or* J*-sectional subspace of* λ. *If* $x_J \in \lambda_J$, *then the canonical preimage of* x_J *is the sequence* $\bar{x}_J$ *which agrees with* x_J *on the indices in* J *and is zero elsewhere. Then,* λ *is called monotone provided* λ *contains the canonical preimages of all its stepspaces.*

It is obvious that the inclusions $\ell_p \subset c_0 \subset c \subset f \subset \ell_\infty$ hold. It follows from

$$|x_k| \le \sup_{k\in\mathbb{N}} |x_k| = \|x\|_\infty \text{ and } |x_k| \le \left(\sum_k |x_k|^p\right)^{1/p} = \|x\|_p$$

that $\lim_{n\to\infty} x^{(n)} = x$ in any of the spaces ℓ_∞, c, c_0 and ℓ_p with $p \geq 1$ implying $\lim_{n\to\infty} x_k^{(n)} = x_k$ for all $k \in \mathbb{N}$. Thus norm convergence in the classical spaces ℓ_∞, c, c_0 and ℓ_p with $p \geq 1$ implies coordinatewise convergence. Since convergence in (ω, d_ω) and coordinatewise convergence are equivalent, Banach spaces ℓ_∞, c, c_0 and ℓ_p with $p \geq 1$ have the property that the metrics obtained by their norms are stronger than the metrics of ω on them. Hence the spaces ℓ_∞, c, c_0 and ℓ_p with $p \geq 1$ are BK-spaces with their natural norms.

We now state the following Lemma 1.1.8- Lemma 1.1.9, Theorem 1.1.10-Theorem 1.1.21 and Corollary 1.1.16 without proofs. The proofs can be obtained from pages 21-32 of Başar [21].

Lemma 1.1.8. *The space bs is norm isomorphic to the space ℓ_∞, i.e., $bs \cong \ell_\infty$.*

1.1.12 The Dual Spaces of a Sequence Space

In this subsection, we focus on the $\alpha-$, $\beta-$, $\gamma-$ and continuous duals of the classical sequence spaces. The multiplier space $S(\lambda, \mu)$ of the sequence spaces λ and μ is defined by

$$S(\lambda, \mu) := \{z = (z_k) \in \omega : xz = (x_k z_k) \in \mu \text{ for all } x = (x_k) \in \lambda\}. \tag{1.4}$$

One can easily observe for a sequence space ν that the inclusions

$$S(\lambda, \mu) \subset S(\nu, \mu) \text{ if } \nu \subset \lambda \text{ and } S(\lambda, \mu) \subset S(\lambda, \nu) \text{ if } \mu \subset \nu \tag{1.5}$$

hold. With the notation of (1.4), the $\alpha-$, $\beta-$*and* $\gamma-$*duals of a sequence space* λ, which are denoted by λ^α, λ^β and λ^γ, respectively, are defined by

$$\lambda^\alpha = S(\lambda, \ell_1), \quad \lambda^\beta = S(\lambda, cs) \quad \text{and} \quad \lambda^\gamma = S(\lambda, bs).$$

The $\alpha-$dual, $\beta-$dual and $\gamma-$dual are also referred to as *Köthe-Toeplitz dual*, *generalized Köthe-Toeplitz dual* and *Garling dual*, respectively. The *continuous dual* λ^* of a sequence space λ is defined as the set of all bounded linear functionals on the space λ.

Now, we give the following lemma and a theorem which shows the connection of the $\beta-$ and continuous duals of FK-spaces:

Lemma 1.1.9. (cf. [97, Remark 1.16, p. 152]) *Let $X \supset \phi$ be an FK-space and $a = (a_k)$ be a given sequence. If the series $\sum_k a_k x_k$ converges for every $x = (x_k) \in X$ then the linear functional $f_a : X \to \mathbb{C}$ defined by $f_a(x) = \sum_k a_k x_k$ is continuous for all $x \in X$.*

Theorem 1.1.10. [135, Theorem 7.2.9, p. 107] *Let $X \supset \phi$ be an FK-space.*

(a) Then there is a linear one to one map $\hat{}: X^\beta \to X^*$.

(b) If X has AK then the map $\hat{}: X^\beta \to X^*$ *is an isomorphism.*

Now we determine the $\alpha-$, $\beta-$and $\gamma-$duals of the classical sequence spaces, and the continuous duals of the spaces ℓ_p, c_0 and c, where $1 \leq p < \infty$.

Theorem 1.1.11. *The following statements hold:*

(i) $\ell_\infty^\alpha = c^\alpha = c_0^\alpha = \ell_1$.

(ii) $\ell_1^\alpha = \ell_\infty$.

Theorem 1.1.12. *Let* $1 < p, q < \infty$ *with* $p^{-1} + q^{-1} = 1$. *Then,* $\ell_p^\alpha = \ell_q$.

Now, we give the following lemma concerning the relation between the $\alpha-$, $\beta-$ and $\gamma-$duals of a solid sequence space:

Lemma 1.1.13. (cf. [35, Theorem 7.1.10 (b), p. 343]) *If* X *is solid, then* $X^\alpha = X^\beta = X^\gamma$.

One can directly derive the $\beta-$ and $\gamma-$duals of the spaces ℓ_∞, c_0, ℓ_p, ω and ϕ which are solid, by combining the results of Theorem 1.1.11, Theorem 1.1.12 and Remark 1.1.18 with Lemma 1.1.13. However, their detailed proof is available in Başar [21].

Theorem 1.1.14. [134, Examples 6.2–6.5, pp. 91–92] *The following statements hold:*

(a) $\ell_p^\beta = \ell_q$ *for all* p *with* $1 \leq p < \infty$ *and* $\ell_\infty^\beta = c^\beta = c_0^\beta = \ell_1$.

(b) Let $1 \leq p < \infty$. *Then the continuous dual* ℓ_p^* *of* ℓ_p *is norm isomorphic to* ℓ_q; *this means* $f \in \ell_p^*$ *if and only if* $f(x) = \sum_k a_k x_k$, $(x \in \ell_p)$, *for some sequence* $a = (a_k)_{k\in\mathbb{N}} \in \ell_q$, *and* $\|f\| = \|a\|_q$.

(c) The continuous dual c_0^* *of* c_0 *is norm isomorphic to* ℓ_1.

(d) We have $f \in c^*$ *if and only if*

$$f(x) = \chi_f \lim_{k\to\infty} x_k + \sum_k a_k x_k, \quad x = (x_k) \in c \tag{1.6}$$

with $a = (a_k) = \{f(e^{(k)})\}_{k\in\mathbb{N}} \in \ell_1$, *where*

$$\chi_f = f(e) - \sum_k f(e^{(k)}), \tag{1.7}$$

$$\|f\| = |\chi_f| + \|a\|_1. \tag{1.8}$$

We remark that the continuous dual of ℓ_∞ is not given by a sequence space.

Now, we give the following theorem without proof related to the $\gamma-$duals of the classical sequence spaces since one can prove each part on the same lines as $\alpha-$ and $\beta-$duals:

Theorem 1.1.15. *The following statements hold:*

(i) $\ell_\infty^\gamma = c^\gamma = c_0^\gamma = \ell_1$.

(ii) $\ell_1^\gamma = \ell_\infty$.

(iii) $\ell_p^\gamma = \ell_q$.

Now, we give the following easy result about the $\alpha-$, $\beta-$ and $\gamma-$duals of the spaces f_0 and f:

Corollary 1.1.16. (see [35, (d) of Theorem 7.1.11] and [29, Proposition 4.5]) *The $\alpha-$, $\beta-$ and $\gamma-$duals of the spaces f_0 and f are the space ℓ_1.*

Theorem 1.1.17. (see [35, (a) of Theorem 7.1.12] and [93, Example 3.11]) $cs^\alpha = bv^\alpha = bv_0^\alpha = bs^\alpha = \ell_1$.

Remark 1.1.18. *Note that $\omega^\alpha = \phi$ and $\phi^\alpha = \omega$.*

Theorem 1.1.19. *The following statements hold:*

(i) $cs^\beta = bv$ (cf. [135, (v) of Theorem 7.3.5]).

(ii) $bv^\beta = cs$ (cf. [135, (iii) of Theorem 7.3.5]).

(iii) $bv_0^\beta = bs$ (cf. [135, (ii) of Theorem 7.3.5]).

(iv) $bs^\beta = bv_0$ (cf. [135, (vi) of Theorem 7.3.5]).

Theorem 1.1.20. *The following statements hold:*

(i) $bs^\gamma = cs^\gamma = bv$ (cf. [135, (v) and (iii) of Theorem 7.3.5]).

(ii) $bv^\gamma = bv_0^\gamma = bs$ (cf. [135, (iv) and (ii) of Theorem 7.3.5]).

Finally, we state the following theorem which yields the continuous dual of the space cs given by Coşkun and Başar [42]:

Theorem 1.1.21. *The continuous dual cs^* of cs is norm isomorphic to bv.*

1.1.13 Paranormed Sequence Spaces

Following Malkowsky and Başar [94], we present a survey of most of the known fundamental results involving the sequence spaces $\ell(p)$, $c_0(p)$, $c(p)$ and $\ell_\infty(p)$, $w_0(p)$, $w(p)$ and $w_\infty(p)$ and $f_0(p)$, $f(p)$. These spaces are generalizations of the classical BK-spaces ℓ_p, c_0, c and ℓ_∞, the spaces w_0^p, w^p and w_∞^p, and f_0, f of sequences that are strongly summable to zero, strongly summable and strongly bounded with index p by the Cesàro method of order 1, and of almost null and almost convergent sequences, respectively. The results include the basic topological properties of the generalized spaces, the complete lists of their known $\alpha-$, $\beta-$, $\gamma-$, functional and continuous duals, and the characterizations of many classes of matrix transformations between them, in particular, the complete list of characterizations of matrix transformations between the spaces $\ell(p)$, $c_0(p)$, $c(p)$ and $\ell_\infty(p)$. Furthermore, a great number of interesting special cases are included. The presented results cover a period of four decades. They are intended to inspire the increasing number of researchers working in related topics, and to provide them with a comprehensive collection of results they may find useful for their work.

Firstly, we give the definition of the concept of paranorm.

Definition 1.1.22. *Let X be a real or complex linear space and g be a function from X to the set $\mathbb{R}$ of real numbers. Then, the pair (X, g) is called a paranormed space and g is a paranorm for X, if the following axioms are satisfied for all elements $x, y \in X$ and for all scalars α:*

(PN.1) $g(x) = 0$ *if* $x = \theta$.

(PN.2) $g(-x) = g(x)$.

(PN.3) $g(x + y) \leq g(x) + g(y)$.

(PN.4) If (α_n) is a sequence of scalars with $\alpha_n \to \alpha$, as $n \to \infty$ and $x_n, x \in X$ for all $n \in \mathbb{N}$ with $x_n \to x$ as $n \to \infty$ then $\alpha_n x_n \to \alpha x$ as $n \to \infty$, in the sense that $g(\alpha_n x_n - \alpha x) \to 0$, as $n \to \infty$.

A paranorm g is said to be total, *if $g(x) = 0$ implies $x = \theta$.*

Assume here and after that $p = (p_k)$ is a bounded sequence of strictly positive real numbers, so that $0 < p_k \leq \sup p_k = H < \infty$ and $M = \max\{1, H\}$. We shall assume throughout that $p_k^{-1} + (p_k')^{-1} = 1$ provided $1 < p_k$ for all $k \in \mathbb{N}$. Now, we may give the spaces $\ell_\infty(p)$, $c(p)$, $c_0(p)$, $\ell(p)$, $\omega_\infty(p)$, $\omega(p)$, $\omega_0(p)$, $f_0(p)$, $f(p)$, $bs(p)$, $cs(p)$ and $cs_0(p)$ as well-known examples of the paranormed sequence spaces:

1.1.14 The Spaces $\ell_\infty(p)$, $c(p)$ and $c_0(p)$

The spaces $\ell_\infty(p)$, $c(p)$ and $c_0(p)$ were defined by Maddox [87] (see also Simons [122]), as follows:

$$\begin{aligned}
\ell_\infty(p) &:= \left\{x = (x_k) \in \omega : \sup_{k \in \mathbb{N}} |x_k|^{p_k} < \infty\right\}, \\
c(p) &:= \left\{x = (x_k) \in \omega : \lim_{k \to \infty} |x_k - l|^{p_k} = 0 \text{ for some } l \in \mathbb{C}\right\}, \\
c_0(p) &:= \left\{x = (x_k) \in \omega : \lim_{k \to \infty} |x_k|^{p_k} = 0\right\}.
\end{aligned}$$

It is a routine verification that each of the sets $\ell_\infty(p)$, $c(p)$ and $c_0(p)$ forms a linear space with respect to the coordinatewise addition and scalar multiplication of sequences. Define the function g_1 on the spaces $\ell_\infty(p)$, $c(p)$ and

$c_0(p)$ by

$$g_1(x) = \sup_{k\in\mathbb{N}} |x_k|^{p_k/M}.$$

Then, $c_0(p)$ and $c(p)$ are complete paranormed spaces paranormed by g_1 if $p \in \ell_\infty$ (cf. [86, Theorem 6]). It is known from [87] that the inclusion $c_0(p) \subset c_0(q)$ holds if and only if $\liminf q_k/p_k > 0$. $\ell_\infty(p)$ is also a complete paranormed space by g_1 if and only if $\inf p_k > 0$.

1.1.15 The Space $\ell(p)$

The space $\ell(p)$ was defined by Maddox [84] (see also Nakano [106]), as follows:

$$\ell(p) := \left\{ x = (x_k) \in \omega : \sum_k |x_k|^{p_k} < \infty \right\}, \quad (0 < p_k < \infty).$$

One can show that the set $\ell(p)$ is a linear space with respect to the coordinatewise addition and scalar multiplication of sequences. Maddox [87] showed that the inclusion $\ell(p) \subset \ell(q)$ holds when $0 < p_k \leq q_k$ for all $k \in \mathbb{N}$. We note here that no restriction such as boundedness has to be place on the sequences (p_k) and (q_k) for the validity of the inclusion. Define the function g_2 on the space $\ell(p)$ by

$$g_2(x) = \left(\sum_k |x_k|^{p_k} \right)^{1/M}.$$

$\ell(p)$ is a complete paranormed space paranormed by g_2 and $\{e^{(k)}\}_{k\in\mathbb{N}}$ is a basis in the space $\ell(p)$.

1.1.16 The Spaces $\omega_\infty(p)$, $\omega(p)$ and $\omega_0(p)$

The spaces $\omega_\infty(p)$, $\omega(p)$ and $\omega_0(p)$ are defined (cf. [87]) by

$$\begin{aligned}
\omega_\infty(p) &:= \left\{ x = (x_k) \in \omega : \sup_{n\in\mathbb{N}} \frac{1}{n+1} \sum_{k=0}^{n} |x_k|^{p_k} < \infty \right\}, \\
\omega(p) &:= \left\{ x = (x_k) \in \omega : \lim_{n\to\infty} \frac{1}{n+1} \sum_{k=0}^{n} |x_k - l|^{p_k} = 0 \text{ for some } l \in \mathbb{C} \right\}, \\
\omega_0(p) &:= \left\{ x = (x_k) \in \omega : \lim_{n\to\infty} \frac{1}{n+1} \sum_{k=0}^{n} |x_k|^{p_k} = 0 \right\}.
\end{aligned}$$

It is known that the sets $\omega_\infty(p)$, $\omega(p)$ and $\omega_0(p)$ are the linear spaces with respect to the coordinatewise addition and scalar multiplication of sequences.

Maddox proved in [86] that the inclusion $\omega_0(p) \subset \omega_0(q)$ holds if and only if $\omega_0(p/q) \subset \omega_0(e)$, where p/q denotes the sequence (p_k/q_k). Define the function g_3 on the spaces $\omega_\infty(p)$, $\omega(p)$ and $\omega_0(p)$ by

$$g_3(x) = \sup_{n\in\mathbb{N}} \left(\frac{1}{n+1} \sum_{k=0}^{n} |x_k|^{p_k} \right)^{1/M}.$$

$\omega(p)$ is a complete paranormed space with g_3 (cf. [86, Theorem 6]) and $\omega_\infty(p)$ is paranormed with g_3 if and only if $0 < \inf\ p_k \leq \sup\ p_k < \infty$ (cf. [85, Theorem 2]).

1.1.17 The Paranormed Space of Almost Convergent Sequences

In this subsection, the indices of sequences start with 0. Again we assume that $p = (p_k)_{k=0}^\infty$ is a bounded sequence of positive real numbers, and $M = \max\{1, \sup_{k\in\mathbb{N}} p_k\}$.

S. Nanda [111] generalized the spaces f_0 and f to the spaces $f_0(p)$ and $f(p)$, respectively, as follows:

$$f_0(p) := \left\{ x = (x_k) \in \omega : \lim_{m\to\infty} |t_{mn}(x)|^{p_m} = 0 \text{ uniformly in } n \right\},$$

$$f(p) := \left\{ x = (x_k) \in \omega : \exists l \in \mathbb{C} \ni \lim_{m\to\infty} |t_{mn}(x) - l|^{p_m} = 0 \text{ uniformly in } n \right\}.$$

The following set inclusions between the spaces $c_0(p)$, $f_0(p)$, $c(p)$ and $f(p)$ hold.

Theorem 1.1.23. ([111, Propositions 1 and 2]) *The following inclusions hold:*

(a) $c_0(p) \subset f_0(p)$, $c(p) \subset f(p)$ *and* $f_0(p) \subset f(p)$.

(b) If $0 < p_m \leq q_m < \infty$, *then* $f_0(p) \subset f_0(q)$ *and* $f(p) \subset f(q)$.

The following results are known concerning the topological structures of the spaces $f_0(p)$ and $f(p)$.

Theorem 1.1.24. ([111, Theorem 1]) *The space* $f_0(p)$ *is a complete linear topological space with respect to the paranorm* g_4 *defined by*

$$g_4(x) = \sup_{m,n\in\mathbb{N}} |t_{m,n}(x)|^{p_m/M}.$$

If $\inf_{m\in\mathbb{N}} p_m > 0$, *then* $f(p)$ *is a complete linear topological space with respect to the paranorm* g_4.

We recall that for $r > 0$, a non-empty subset Y of a linear space X is said to be absolutely r-convex if $x, y \in Y$ and $\alpha, \beta \in \mathbb{C}$ with $|\alpha|^r + |\beta|^r \leq 1$ together implying that $\alpha x + \beta y \in Y$. A linear topological space X is said to be r-convex (cf. [89]) if every neighbourhood of $\theta \in X$ contains an absolutely r-convex neighbourhood of $\theta \in X$. The next result is known.

Theorem 1.1.25. ([111, Proposition 3]) *The spaces $f_0(p)$ and $f(p)$ are 1-convex.*

Several classes of matrix transformations into the normed or paranormed spaces of almost convergent or almost null sequences were characterized in [111, 108]. Let $A = (a_{nk})_{n,k=0}^{\infty}$ be an infinite matrix. Then we write

$$a(n,k,m) = \frac{1}{m+1}\sum_{j=0}^{m} a_{n+j,k} \text{ for all } n,k,m \in \mathbb{N}.$$

Theorem 1.1.26. (cf. [111, Theorems 2, 3, 5, 6]) *We have*

(a) $A \in (c_0(p) : f_0(p))$ if and only if

(i) there exists an integer $B > 1$ such that

$$C_n = \sup_{m\in\mathbb{N}}\left(\sum_k |a(n,k,m)|B^{-1/p_k}\right)^{p_m} < \infty \text{ for all } n, \tag{1.9}$$

$$\lim_{m\to\infty} |a(n,k,m)|^{p_m} \text{ uniformly in } n. \tag{1.10}$$

(b) $A \in (c(p) : f)$ if and only if there exists an integer $B > 1$ such that

$$D_n = \sup_{m\in\mathbb{N}}\sum_k |a(n,k,m)|B^{-1/p_k} < \infty \text{ for all } n, \tag{1.11}$$

$$\lim_{m\to\infty} a(n,k,m) = \alpha_k \text{ uniformly in } n \text{ for each } k, \tag{1.12}$$

$$\lim_{m\to\infty}\sum_k a(n,k,m) = \alpha \text{ uniformly in } n. \tag{1.13}$$

(c) $A \in (\ell_\infty(p) : f)$ if and only if the condition (1.12) is satisfied and for all integers $B > 1$

$$\sup_{m\in\mathbb{N}}\sum_k |a(n,k,m)| < \infty \text{ for all } n, \tag{1.14}$$

$$\lim_{m\to\infty} |a(n,k,m) - \alpha_k|B^{1/p_k} = 0 \text{ uniformly in } n. \tag{1.15}$$

(d) $A \in (\ell_1 : f_0)$ if and only if

$$\sup_{n,k\in\mathbb{N}} |a(n,k,m)| < \infty \text{ for all } n, \tag{1.16}$$

$$\lim_{m\to\infty} a(n,k,m) = 0 \text{ uniformly in } n. \tag{1.17}$$

Remark 1.1.27. *The following statements hold:*

(a) *The characterizations of the classes $(c_0(p) : f)$ and $(\ell_\infty(p) : f_0)$ can be found in* [111, Corollaries p. 180 and 181] *as special cases of* [111, Theorems 3 and 5].

(b) *Parts (b) and (c) of Theorem 1.1.26 generalize the characterizations of the classes* $(c : f)$ *and* $(\ell_\infty : f)$ *established in* [71, Theorem 3.1] *and* [55, Theorem 2.1].

We close with the characterizations of the classes $(\ell(p) : f)$ and $(w(p) : f)$ in [108, Theorems 1 and 5].

Theorem 1.1.28. *We have the following:*

(a) $A \in (\ell(p) : f)$ *if and only if the condition (1.12) in Part (b) of Theorem 1.1.26 holds and there exists an integer* $B > 1$ *such that for all* n

$$\sup_{m\in\mathbb{N}} \sum_k |a(n,k,m)|^{q_k} B^{-q_k} < \infty \text{ if } p_k < 1 \text{ and } q_k = p_k/(p_k - 1), \quad (1.18)$$

$$\sup_{m,k\in\mathbb{N}} |a(n,k,m)|^{p_k} < \infty \quad \text{if} \quad 0 < p_k \le 1. \quad (1.19)$$

(b) *If* $0 < p_k \le 1$ *for all* $k \in \mathbb{N}$, *then* $A \in (w(p) : f)$ *if and only if the conditions (1.12) and (1.13) in Part (b) of Theorem 1.1.28 hold, and there exists* $B > 1$ *such that*

$$C_n = \sup_{m\in\mathbb{N}} \sum_\nu \max_\nu \left(\left(2^\nu B^{-1}\right)^{1/p_k} |a(n,k,m)| \right) < \infty \quad (1.20)$$

for all $n \in \mathbb{N}$.

1.1.18 The Spaces $bs(p)$, $cs(p)$ and $cs_0(p)$

The spaces $bs(p)$, $cs(p)$ and $cs_0(p)$ are defined by

$$bs(p) := \left\{ x = (x_k) \in \omega : \sup_{k\in\mathbb{N}} \left| \sum_{j=0}^k x_j \right|^{p_k} < \infty \right\},$$

$$cs(p) := \left\{ x = (x_k) \in \omega : \lim_{k\to\infty} \left| \sum_{j=0}^k x_j - l \right|^{p_k} = 0 \text{ for some } l \in \mathbb{C} \right\},$$

$$cs_0(p) := \left\{ x = (x_k) \in \omega : \lim_{k\to\infty} \left| \sum_{j=0}^k x_j \right|^{p_k} = 0 \right\}.$$

It is not hard to establish that each of the sets $bs(p)$, $cs(p)$ and $cs_0(p)$ forms a linear space with respect to the coordinatewise addition and scalar multiplication of sequences.

Define the function g_5 on the spaces $bs(p)$, $cs(p)$ and $cs_0(p)$ by

$$g_5(x) = \sup_{k\in\mathbb{N}} \left| \sum_{j=0}^k x_j \right|^{p_k/M}.$$

$bs(p)$, $cs(p)$ and $cs_0(p)$ are the complete paranormed spaces by g_5 if and only if $\inf p_k > 0$.

In general $\ell_\infty(p)$, $c(p)$, $c_0(p)$, $\ell(p)$, $\omega_\infty(p)$, $\omega(p)$, $\omega_0(p)$, $bs(p)$, $cs(p)$ and $cs_0(p)$ are not normed spaces. If we choose $p_k = p$ for all $k \in \mathbb{N}$, then the paranormed spaces $\ell_\infty(p)$, $c(p)$, $c_0(p)$, $\ell(p)$, $\omega_\infty(p)$, $\omega(p)$, $\omega_0(p)$, $bs(p)$, $cs(p)$ and $cs_0(p)$ are reduced to Banach spaces ℓ_∞, c, c_0, ℓ_p, ω_∞^p, ω^p, ω_0^p, bs, cs and cs_0, respectively.

1.2 Matrix Domains in Sequence Spaces

This section is devoted to the domains of some triangle matrices, with a special emphasis on the Nörlund, composition of the first order Cesàro, Euler and backward difference matrices and the double sequential band matrix. Also, the Schauder bases of those spaces, their $\alpha-$, $\beta-$ and $\gamma-$duals, and the characterizations of some classes of matrix transformations are given.

1.2.1 Preliminaries, Background and Notations

Let λ, μ be any two sequence spaces and $A = (a_{nk})$ be an infinite matrix of complex numbers a_{nk}, where $k, n \in \mathbb{N}$. Then, we say that A defines a *matrix transformation* from λ into μ and we denote it by writing $A : \lambda \to \mu$, if for every sequence $x = (x_k) \in \lambda$ the sequence $Ax = \{(Ax)_n\}$, the A-transform of x, is in μ, where

$$(Ax)_n = \sum_k a_{nk}x_k \tag{1.1}$$

provided the series on the right side of (1.1) converges for each $n \in \mathbb{N}$. By $(\lambda : \mu)$, we denote the class of all matrices A such that $A : \lambda \to \mu$. Thus, $A \in (\lambda : \mu)$ if and only if Ax exists, i.e., $A_n \in \lambda^\beta$ for all $n \in \mathbb{N}$ and belongs to μ for all $x \in \lambda$, where A_n denotes the sequence in the nth row of A.

The *domain* λ_A of an infinite matrix A in a sequence space λ is defined by

$$\lambda_A = \Big\{x = (x_k) \in \omega : Ax \in \lambda\Big\} \tag{1.2}$$

which is a sequence space. If A is a triangle, then one can easily observe that the sequence spaces λ_A and λ are linearly isomorphic, i.e., $\lambda_A \cong \lambda$. If λ is a sequence space, then the continuous dual λ_A^* of the space λ_A is given (cf. [135, Theorem 4.4.2, p. 66]) by $\lambda_A^* = \{f : f = g \circ A,\ g \in \lambda^*\}$. Although in most cases the new sequence space λ_A generated by the matrix A from a sequence space λ is the expansion or the contraction of the original space λ, it may be observed in some cases that those spaces overlap. Indeed, one can easily see that the inclusion $\lambda_S \subset \lambda$ strictly holds for $\lambda \in \{\ell_\infty, c, c_0\}$. As this, one can deduce that the inclusion $\lambda \subset \lambda_\nabla$ also strictly holds for $\lambda \in \{\ell_\infty, c, c_0, \ell_p\}$, where $0 < p < \infty$. However, if we define $\lambda = c_0 \oplus span\{z\}$ with $z = \{(-1)^k\}$, i.e., $x \in \lambda$ if and only if $x = s + \alpha z$ for some $s \in c_0$ and some $\alpha \in \mathbb{C}$, and consider the matrix A with the rows A_n defined by $A_n = (-1)^n e^{(n)}$ for all $n \in \mathbb{N}$, we have $Ae = z \in \lambda$ but $Az = e \notin \lambda$ which lead us to the consequences that $z \in \lambda \setminus \lambda_A$ and $e \in \lambda_A \setminus \lambda$. That is to say that the sequence spaces λ_A and λ overlap but neither contains the other.

The approach constructing a new sequence space by means of the matrix domain of a triangle has recently been employed by Wang [133], Ng and Lee [115], Malkowsky [92], Altay and Başar [4, 5, 6, 7, 8, 9, 10, 11], Aydın and Başar [12, 13, 14, 15, 16, 17, 18], Malkowsky and Savaş [99], Başarır [30, 31], Başarır and Kayıkçı [33], Başarır and Öztürk [34], Kara and Başarır [67], Kara et al. [68], Başar et al. [26], Şengönül and Başar [128], Altay [3], Polat and Başar [120] and Malkowsky et al. [95]. Additionally, $c_0(u,p)$ and $c(u,p)$ are the spaces consisting of the sequences $x = (x_k)$ such that $ux = (u_k x_k)$ is in the spaces $c_0(p)$ and $c(p)$ for $u \in \mathcal{U}$, respectively, and studied by Başarır [30]. Finally, the new technique for deducing certain topological properties, for example AB-, KB-, AD-properties, solidity and monotonicity, etc., and determining the $\beta-$ and $\gamma-$duals of the domain of a triangle matrix in a sequence space is given by Altay and Başar [9].

Prior to giving the theorem on the Schauder basis of the matrix domain of a triangle in a linear metric space with a Schauder basis, we record the following result concerning the unique inverse of a triangle:

Proposition 1.2.1. ([135, 1.4.8, p. 9] and [41, Remark 22 (a), p. 22]) *Every triangle U has a unique inverse V which is also a triangle and $x = U(Vx) = V(Ux)$ for all $x \in \omega$.*

Throughout this section, let U always denote a triangle and V be its inverse. The reader may refer to [98] for the general results on the Schauder bases of matrix domains of triangles in FK-spaces and the β–duals of such matrix domains in FK-spaces with AK, and in ℓ_∞ and c, and the characterizations of matrix transformations on such spaces. The following general result technically reduces the determination of a Schauder basis of the matrix domain of a triangle in a linear metric space with a Schauder basis to the determination of the inverse of the triangle:

Theorem 1.2.2. [61, Theorem 2.3] *If $\{b(n)\}_{n\in\mathbb{N}}$ is a Schauder basis of the linear metric sequence space (X,d), then $\{V(b^{(n)})\}_{n\in\mathbb{N}}$ is a basis of $Z = X_U$ with the metric d_U defined by $d_U(z,\tilde{z}) = d(Uz, U\tilde{z})$ for all $z, \tilde{z} \in Z$.*

Proof. It is clear that d_U is a metric on $Z = X_U$. We write $c^{(n)} = Vb^{(n)}$ for $n \in \mathbb{N}$. First, we note that $c^{(n)} \in Z$ for all $n \in \mathbb{N}$, since $Uc^{(n)} = U(Vb^{(n)}) = b^{(n)}$ by Proposition 1.2.1, and $b^{(n)} \in X$.

Let $z \in Z$ be given. Then $x = Uz \in X$ and there is a unique sequence $(\lambda_n)_{n\in\mathbb{N}}$ of scalars such that $x^{\langle m\rangle} = \sum_{n=0}^{m} \lambda_n b^{(n)} \to x$, as $m \to \infty$. We put

$x^{\langle m\rangle} = \sum_{n=0}^{m} \lambda_n c^{(n)}$ for $m \in \mathbb{N}$. Then, it follows that

$$U\left(x^{\langle m\rangle}\right) = \sum_{n=0}^{m} \lambda_n U\left(c^{(n)}\right) = \sum_{n=0}^{m} \lambda_n b^{(n)} = x^{\langle m\rangle}$$

for all $m \in \mathbb{N}$, hence

$$\lim_{m\to\infty} d_U\left(z^{\langle m\rangle}, z\right) = \lim_{m\to\infty} d\left(Uz^{\langle m\rangle}, Uz\right) \sum_{n=0}^{m} \lambda_n U\left(c^{(n)}\right) = \lim_{m\to\infty} d\left(x^{\langle m\rangle}, x\right) = 0,$$

as desired. □

Since $X = (X_U)_V$ by Proposition 1.2.1, an application of Theorem 1.2.2 leads to the following:

Remark 1.2.3. *The matrix domain X_U of a linear metric sequence space has a basis if and only if X has a basis.*

Now, we give the Schauder bases of FK-spaces with AK.

Corollary 1.2.4. [61, Corollary 2.5] *Let X be an FK-space with AK and the sequences $c^{(n)} = \left\{c_k^{(n)}\right\}$ and $c^{(-1)} = \left\{c_k^{(-1)}\right\}$ be defined by*

$$c_k^{(n)} = \begin{cases} 0 & , \quad 0 \le k \le n-1, \\ v_{kn} & , \quad k \ge n, \end{cases} \qquad \text{and} \quad c_k^{(-1)} = \sum_{j=0}^{k} v_{kj}$$

for all $k \in \mathbb{N}$. Then the following statements hold:

(a) Every sequence $z = (z_n)_{n\in\mathbb{N}} \in Z = X_U$ has a unique representation

$$z = \sum_n (Uz)_n c^{(n)}. \tag{1.3}$$

(b) Every sequence $v = (v_n)_{n\in\mathbb{N}} \in Y = X_U \oplus e$ has a unique representation

$$v = le + \sum_n \{U(v - le)\}_n,$$

where $l \in \mathbb{C}$ is uniquely determined such that $v = z + le$ for $z \in Z = X_U$.

(c) Every sequence $t = (t_n)_{n\in\mathbb{N}} \in T = (X \oplus e)_U$ has a unique representation

$$t = lc^{(-1)} + \sum_n [(Ut)_n - l]c^{(n)}, \tag{1.4}$$

where $l \in \mathbb{C}$ is uniquely determined such that $Ut - le \in X$.

Proof. First, we note that $c^{(n)} = Ve^{(n)}$ for all $n \in \mathbb{N}$ and $c^{(-1)} = Ve$; hence the sequences $\{c^{(n)}\}_{n\in\mathbb{N}}$ and $\{c^{(n)}\}_{n\in\mathbb{N}_{-1}}$ are bases of Z and T, respectively, by Theorem 1.2.2.

(a) Let $z = (z_n)_{n\in\mathbb{N}} \in Z$ be given. Then $x = Uz \in X$ and (1.3) follows if we take $\lambda_n = (Uz)_n$ for all $n \in \mathbb{N}$ in the proof of Theorem 1.2.2.

(b) Let $v = (v_n)_{n\in\mathbb{N}} \in Y = X_U \oplus e$ be given. Then there is a uniquely determined sequence $z \in Z$ and $l \in \mathbb{C}$ such that $v = z + le$ and we have $z = \sum_n (Uz)_n c^{(n)}$ by Part (a). It follows that $v = le+z = le+\sum_n \{U(v-le)\}_n c^{(n)}$.

(c) Let $t = (t_n)_{n\in\mathbb{N}} \in T$. Then, $u = Ut \in X \oplus e$ and there are uniquely determined $x \in X$ and $l \in \mathbb{C}$ such that $u = x + le$. We put $z = t - lc^{(-1)}$. Then, $z \in Z$, since $Uz = U(t - lc^{(-1)}) = U(t - lc^{(-1)}) = Ut - lUc^{(-1)} = u - le = x \in X$, and so we have $z = \sum_n (Uz)_n c^{(n)} = \sum_n [(Ut)_n - l] c^{(n)}$ by Part (a). Now (1.4) is an immediate consequence, since $t = z + lc^{(-1)}$. □

Throughout this chapter, we need the following triangles. We write $\mathcal{U}$ for the set of all sequences $u = (u_k)$ such that $u_k \neq 0$ for all $k \in \mathbb{N}$. For $u \in \mathcal{U}$, let $1/u = (1/u_k)$. Let $u, v, z \in \mathcal{U}$ and define the summation matrix $S = (s_{nk})$, the backward difference matrix $\nabla = (d_{nk})$, the Cesàro matrix $C = (c_{nk})$ of order one, the Riesz matrix $R^q = (r^q_{nk})$ with respect to the sequence $q = (q_k)$, the matrix $A^r = \{a^r_{nk}\}$, the Euler matrix E^r of order r by

$$s_{nk} = \begin{cases} 1 & , \quad 0 \leq k \leq n, \\ 0 & , \quad k > n, \end{cases} \qquad d_{nk} = \begin{cases} (-1)^{n-k} & , \quad n-1 \leq k \leq n, \\ 0 & , \quad \text{otherwise}, \end{cases}$$

$$c_{nk} = \begin{cases} \frac{1}{n+1} & , \quad 0 \leq k \leq n, \\ 0 & , \quad k > n, \end{cases} \qquad r^q_{nk} = \begin{cases} \frac{q_k}{Q_n} & , \quad 0 \leq k \leq n, \\ 0 & , \quad k > n, \end{cases}$$

$$a^r_{nk} = \begin{cases} \frac{1+r^k}{n+1} & , \quad 0 \leq k \leq n, \\ 0 & , \quad k > n, \end{cases} \qquad e^r_{nk} = \begin{cases} \binom{n}{k}(1-r)^{n-k} r^k & , \quad 0 \leq k \leq n, \\ 0 & , \quad k > n \end{cases}$$

for all $k, n \in \mathbb{N}$, where $q = (q_k)$ is a sequence of non-negative real numbers with $q_0 > 0$ and $Q_n = \sum_{k=0}^n q_k$ for all $n \in \mathbb{N}$, and $0 < r < 1$.

Chapter 2

Some Normed Sequence Spaces Generated by Certain Triangles

Abstract. *In this chapter, we emphasize on the domains of Nörlund means, domains of the Euler-Cesàro difference matrix and domains of the band matrix* $\widehat{F}$ *defined by the Fibonacci sequence in the classical sequence spaces.*

Keywords. *Domain of an infinite matrix, Nörlund mean, Cesàro mean of order one, Backward difference matrix, Euler mean of order one, the band matrix* $\widehat{F}$ *defined by the Fibonacci sequence, Schauder basis, alpha-, beta- and gamma-duals, Characterization of the matrix transformations.*

2.1 Normed Nörlund Sequence Spaces

Following Tuğ and Başar [130, 131], we introduce some normed Nörlund spaces.

Now, following Peyerimhoff [119, pp. 17–19] and Mears [101] we give a short background on the Nörlund means. Let $t = (t_k)$ be a sequence of non-negative real numbers with $t_0 > 0$ and write $T_n = \sum_{k=0}^{n} t_k$ for all $n \in \mathbb{N}$. Then, the Nörlund means with respect to the sequence $t = (t_k)$ is defined by the matrix $N^t = (a_{nk}^t)$ which is given by

$$a_{nk}^t = \begin{cases} \dfrac{t_{n-k}}{T_n} & , \quad 0 \leq k \leq n, \\ 0 & , \quad k > n \end{cases}$$

for all $k, n \in \mathbb{N}$. It is known that the Nörlund matrix N^t is a Toeplitz matrix if and only if $t_n/T_n \to 0$, as $n \to \infty$, and is reduced in the case $t = e = (1, 1, 1, \ldots)$ to the matrix C_1 of first order Cesàro mean. Additionally, for $t_n = A_n^{r-1}$ for all $n \in \mathbb{N}$, the method N^t is reduced to the Cesàro method C_r of order $r > -1$, where

$$A_n^r = \begin{cases} \dfrac{(r+1)(r+2)\cdots(r+n)}{n!} & , \quad n = 1, 2, 3, \ldots, \\ 1 & , \quad n = 0. \end{cases}$$

Let $t_0 = D_0 = 1$ and define D_n for $n \in \{1, 2, 3, \ldots\}$ by

$$D_n = \begin{vmatrix} t_1 & 1 & 0 & 0 & \ldots & 0 \\ t_2 & t_1 & 1 & 0 & \ldots & 0 \\ t_3 & t_2 & t_1 & 1 & \ldots & 0 \\ \vdots & \vdots & \vdots & \vdots & \ddots & \vdots \\ t_{n-1} & t_{n-2} & t_{n-3} & t_{n-4} & \ldots & 1 \\ t_n & t_{n-1} & t_{n-2} & t_{n-3} & \ldots & t_1 \end{vmatrix}.$$

The inverse matrix $U^t = (u^t_{nk})$ of the matrix N^t is given by Mears in [101], as follows

$$u^t_{nk} = \begin{cases} (-1)^{n-k} D_{n-k} T_k & , \quad 0 \leq k \leq n, \\ 0 & , \quad k > n \end{cases}$$

for all $k, n \in \mathbb{N}$. Also, one can derive by straightforward calculation for all $k \in \{1, 2, 3, \ldots\}$ that

$$D_k = \sum_{j=1}^{k-1} (-1)^{j-1} t_j D_{k-j} + (-1)^{k-1} t_k.$$

Additionally, the inverse of Nörlund matrix and some multiplication theorems for Nörlund mean were studied by Mears [100, 101].

The domain of Nörlund matrix N^t in the classical sequence spaces ℓ_∞ and ℓ_p were introduced by Wang [133], where $1 \leq p < \infty$. We should note here that as a new development, the reader may refer to [43] for studying about sequence spaces and related topics in the sense of multiplicative calculus.

We define the sequence $y = (y_k)$ by the N^t-transform of a sequence $x = (x_k)$, that is,

$$y_k = (N^t x)_k = \frac{1}{T_k} \sum_{j=0}^{k} t_{k-j} x_j \tag{2.1}$$

for all $k \in \mathbb{N}$. Therefore, by applying U^t to the sequence y defined by (2.1) we obtain that

$$x_k = (U^t y)_k = \sum_{j=0}^{k} (-1)^{k-j} D_{k-j} T_j y_j \tag{2.2}$$

for all $k \in \mathbb{N}$. Throughout the Subsections 2.1.1 and 4.1.1, we suppose that the terms of the sequences $x = (x_k)$ and $y = (y_k)$ are connected with the relation (2.1).

2.1.1 The Sequence Spaces $c_0(N^t)$ and $c(N^t)$ of Non-absolute Type

We introduce the sequence spaces $c_0(N^t)$ and $c(N^t)$ as the set of all sequences whose N^t-transforms are in the spaces of null and convergent sequences, respectively, that is

$$c_0(N^t) := \left\{x = (x_k) \in \omega : \lim_{n\to\infty} \frac{1}{T_n} \sum_{k=0}^{n} t_{n-k} x_k = 0\right\},$$

$$c(N^t) := \left\{x = (x_k) \in \omega : \exists l \in \mathbb{C} \text{ such that } \lim_{n\to\infty} \sum_{k=0}^{n} \frac{t_{n-k}}{T_n} x_k = l\right\}.$$

Theorem 2.1.1. *The sets $c_0(N^t)$ and $c(N^t)$ are the linear spaces with the coordinatewise addition and scalar multiplication which are the BK-spaces with the norm $\|x\|_{c_0(N^t)} = \|x\|_{c(N^t)} = \|N^t x\|_\infty$.*

Proof. The proof of the first part of the theorem is a routine verification and so we omit details.

Since c_0 and c are the BK-spaces with respect to their usual sup-norm and N^t is a triangle matrix, Theorem 4.3.2 of Wilansky [135, p. 61] gives the fact that $c_0(N^t)$ and $c(N^t)$ are the BK-spaces.

This completes the proof. □

Let λ denote any of the spaces c_0 or c. With the notation of (2.1), since the transformation $T : \lambda(N^t) \to \lambda$ defined by $x \mapsto y = Tx = N^t x$ is a norm preserving linear bijection, we have the following:

Corollary 2.1.2. *Let λ denote any of the spaces c_0 and c. Then, the sequence space $\lambda(N^t)$ is linearly norm isomorphic to the space λ.*

Theorem 2.1.3. *Let N^t be a non-Mercerian matrix, i.e., $c_{N^t} \neq c$. Then, the inclusions $c_0 \subset c_0(N^t)$ and $c \subset c(N^t)$ strictly hold.*

Proof. Suppose that N^t is a non-Mercerian matrix. To show the inclusion relation $c_0 \subset c_0(N^t)$ holds we take any sequence $y \in c_0$. Then, by using the regularity property of N^t we can easily find that $N^t y \in c_0$ which means that $y \in c_0(N^t)$. That is to say that the inclusion $c_0 \subset c_0(N^t)$ holds. In a similar way, it is trivial to see that the inclusion $c \subset c(N^t)$ also holds.

To prove the second part of the theorem, we should show that the sets $c_0(N^t) \setminus c_0$ and $c(N^t) \setminus c$ are not empty. For this, consider the sequence $v = (v_k) = \{(-1)^k\}$ which does not belong to both of the spaces c_0 and c. Since

$$\lim_{k\to\infty} (Cv)_k = \lim_{k\to\infty} \frac{1}{k+1} \sum_{j=0}^{k} (-1)^j = 0$$

and Cesàro mean C of order one is a special case of Nörlund mean, we have $v \in c_0(N^t)$. Additionally, since $Cv \in c_0$ implies $Cv \in c$, we also have $v \in$

$c(N^t)$. Hence, $v \in [c_0(N^t) \setminus c_0] \cap [c(N^t) \setminus c]$. That is to say that the inclusions $c_0 \subset c_0(N^t)$ and $c \subset c(N^t)$ strictly hold. □

Remark 2.1.4. [60, Remark 2.4] The domain λ_T of an infinite matrix $T = (t_{nk})$ in a normed sequence space λ has a basis if and only if λ has a basis, if T is a triangle.

As a direct consequence of Remark 2.1.4, we have:

Corollary 2.1.5. *Let $\alpha_k = (N^t x)_k$ for all $k \in \mathbb{N}$. Define the sequence $u^{(n)} = \left\{u_k^{(n)}\right\}_{k\in\mathbb{N}}$ in the space $c_0(N^t)$ by*

$$u_k^{(n)} = \begin{cases} (-1)^{n-k} D_{n-k} T_k & , \quad 0 \le k \le n, \\ \\ 0 & , \quad k > n \end{cases} \tag{2.3}$$

for every fixed $n \in \mathbb{N}$.

(a) The sequence $\{u^{(n)}\}_{n\in\mathbb{N}}$ is a basis for the space $c_0(N^t)$ and any $x \in c_0(N^t)$ has a unique representation of the form $x = \sum_k \alpha_k u_k^n$.

(b) The set $\{e, u^{(n)}\}$ is a basis for the sequence space $c(N^t)$ and any $x \in c(N^t)$ has a unique representation of the form $x = le + \sum_k (\alpha_k - l) u_k^n$, where $l = \lim_{k\to\infty} \alpha_k$.

2.1.2 The Alpha-, Beta- and Gamma-duals of the Spaces $c_0(N^t)$ and $c(N^t)$

In this subsection, the alpha-, beta- and gamma-duals of the spaces $c_0(N^t)$ and $c(N^t)$ are determined.

Now, we start with the following lemma due to Steieglitz and Tietz [127] which is needed in proving our theorems, below. Here and after, we denote the collection of all finite subsets of $\mathbb{N}$ by $\mathcal{F}$.

Lemma 2.1.6. *Let $A = (a_{nk})$ be an infinite matrix over the complex field. Then, the following statements hold:*

(a) $A \in (c_0 : \ell_1) = (c : \ell_1) = (\ell_\infty : \ell_1)$ if and only if

$$\sup_{K\in\mathcal{F}} \sum_n \left| \sum_{k\in K} a_{nk} \right| < \infty.$$

(b) $A \in (c_0 : \ell_\infty) = (c : \ell_\infty) = (\ell_\infty : \ell_\infty)$ if and only if

$$\sup_{n\in\mathbb{N}} \sum_k |a_{nk}| < \infty. \tag{2.4}$$

(c) $A \in (c : c)$ *if and only if (2.4) holds, and*

$$\exists a_k \in \mathbb{C} \text{ such that } \lim_{n\to\infty} a_{nk} = a_k \text{ for all } k \in \mathbb{N}, \tag{2.5}$$

$$\exists a \in \mathbb{C} \text{ such that } \lim_{n\to\infty} \sum_k a_{nk} = a. \tag{2.6}$$

(d) $A \in (c : c_0)$ *if and only if (2.4) holds, and*

$$\lim_{n\to\infty} a_{nk} = 0 \text{ for all } k \in \mathbb{N}, \tag{2.7}$$

$$\lim_{n\to\infty} \sum_k a_{nk} = 0. \tag{2.8}$$

Theorem 2.1.7. *The α-dual of the spaces $c_0(N^t)$ and $c(N^t)$ is the set*

$$d_1^t := \left\{ a = a_k \in \omega : \sup_{K\in\mathcal{F}} \sum_n \left| \sum_{k\in K} (-1)^{n-k} D_{n-k} T_k a_n \right| < \infty \right\}.$$

Proof. Let us define the matrix $B = (b_{nk}^t)$ with the aid of $a = (a_k) \in \omega$ by

$$b_{nk}^t = \begin{cases} (-1)^{n-k} D_{n-k} T_k a_n & , \quad 0 \le k \le n \\ 0 & , \quad k > n \end{cases} \tag{2.9}$$

for all $k, n \in \mathbb{N}$. Since the relation (2.2) holds, we easily obtain that

$$a_n x_n = \sum_{k=0}^{n} (-1)^{n-k} D_{n-k} T_k a_n y_k = (By)_n \tag{2.10}$$

for all $n \in \mathbb{N}$. Then, we conclude by (2.10) that $ax = (a_n x_n) \in \ell_1$ whenever $x \in c_0(N^t)$ or $\in c(N^t)$ if and only if $By \in \ell_1$ whenever $y \in c_0$ or $\in c$. Therefore, we derive by Part (a) of Lemma 2.1.6 that

$$\sup_{K\in\mathcal{F}} \sum_n \left| \sum_{k\in K} (-1)^{n-k} D_{n-k} T_k a_n \right| < \infty$$

which leads to the desired result that $\{c_0(N^t)\}^\alpha = \{c(N^t)\}^\alpha = d_1^t$. □

Theorem 2.1.8. *Define the set d_2^t, as follows*

$$d_2^t := \left\{ a = (a_k) \in \omega : \sup_{n\in\mathbb{N}} \sum_k \left| \sum_{j=k}^{n} (-1)^{j-k} D_{j-k} T_k a_j \right| < \infty \right\}.$$

Then, $\{c_0(N^t)\}^\beta = \{c(N^t)\}^\beta = d_2^t \cap cs$.

Proof. Let $x = (x_k)$ be in $c_0(N^t)$ or $c(N^t)$. Now, consider the equality

$$\begin{aligned} \sum_{k=0}^{n} a_k x_k &= \sum_{k=0}^{n-1}\sum_{j=k}^{n}(-1)^{j-k}a_j D_{j-k}T_k y_k + a_n T_n y_n \\ &= (Ey)_n \end{aligned} \tag{2.11}$$

for all $n \in \mathbb{N}$, where $E = (e_{nk})$ is defined by

$$e_{nk} = \begin{cases} \sum_{j=k}^{n}(-1)^{j-k}D_{j-k}T_k a_j &, \quad 0 \leq k \leq n-1, \\ a_n T_n &, \quad k = n, \\ 0 &, \quad k > n \end{cases} \tag{2.12}$$

for all $k, n \in \mathbb{N}$. Then, we observe by taking into the equality (2.11) that $ax = (a_k x_k) \in cs$ whenever $x = (x_k) \in c(N^t)$ if and only if $Ey \in c$ whenever $y = (y_k) \in c$. This is equivalent to the statement that "$a = (a_k) \in \{c(N^t)\}^\beta$ if and only if $E \in (c : c)$." Therefore, we derive from (2.11) and Part (c) of Lemma 2.1.6 that the sequence (a_k) satisfies the following conditions, respectively,

$$\sup_{n\in\mathbb{N}} \sum_k \left| \sum_{j=k}^{n}(-1)^{j-k}D_{j-k}T_k a_j \right| < \infty,$$

$$(a_k) \in cs.$$

This show that $\{c(N^t)\}^\beta = d_2^t \cap cs$, as asserted. □

Theorem 2.1.9. *The γ-dual of the spaces $c_0(N^t)$ and $c(N^t)$ is the set d_2^t.*

Proof. This is similar to the proof of Theorem 2.1.8 with Part (b) of Lemma 2.1.6 instead of Part (c) of Lemma 2.1.6. So, we omit details. □

2.1.3 Matrix Transformations Related to the Sequence Space $c(N^t)$

In this subsection, we characterize some classes of matrix transformations from the spaces $c(N^t)$ into the classical sequence spaces ℓ_∞, c and c_0. Additionally, we characterize the class of matrix transformations from a given sequence space μ to the space $c(N^t)$.

Throughout this subsection, we define the matrices $F = (f_{nk})$ and $G = (g_{nk})$ via multiplication of the matrices A and N^t by the products AN^t and $N^t A$, respectively, that is

$$f_{nk} = \sum_{j=k}^{\infty}(-1)^{j-k}D_{j-k}T_k a_{nj} \text{ and } g_{nk} = \sum_{j=0}^{n}\frac{t_{n-j}}{T_n}a_{jk} \tag{2.13}$$

for all $k, n \in \mathbb{N}$.

Theorem 2.1.10. *$A = (a_{nk}) \in (c(N^t) : \ell_\infty)$ if and only if*

$$A_n \in \{c(N^t)\}^\beta \text{ for each } n \in \mathbb{N}, \tag{2.14}$$
$$F \in (c : \ell_\infty). \tag{2.15}$$

Proof. Suppose that $A = (a_{nk}) \in (c(N^t) : \ell_\infty)$ and $x = (x_k) \in c(N^t)$. Consider the following equality derived from the mth partial sum of the series $\sum_k a_{nk}x_k$:

$$\sum_{k=0}^{m} a_{nk}x_k = \sum_{k=0}^{m}\sum_{j=k}^{m}(-1)^{j-k}D_{j-k}a_{nj}T_k y_k \tag{2.16}$$

for all $m, n \in \mathbb{N}$. Since Ax exists and belongs to the space ℓ_∞, the necessity of the condition (2.14) is obvious. Therefore, by letting $m \to \infty$ in the equality (2.16) one can see that

$$\sum_k a_{nk}x_k = \sum_k \sum_{j=k}^{\infty}(-1)^{j-k}D_{j-k}T_k a_{nj}y_k \tag{2.17}$$

for all $n \in \mathbb{N}$, i.e., $Ax = Fy$ which gives that $Fy \in \ell_\infty$. That is to say that $F \in (c : \ell_\infty)$.

Conversely, let us suppose that the conditions (2.14) and (2.15) hold, and take $x = (x_k) \in c(N^t)$. Then, (2.14) implies the existence of Ax and since the spaces $c(N^t)$ and c are isomorphic we have $y \in c$. Therefore, (2.17) gives with (2.4) with f_{nk} instead of a_{nk} that

$$\begin{aligned} \|Ax\|_\infty &= \sup_{n\in\mathbb{N}}\left|\sum_k a_{nk}x_k\right| \\ &\leq \sup_{n\in\mathbb{N}}\sum_k\left|\sum_{j=k}^{\infty}(-1)^{j-k}D_{j-k}T_k a_{nj}y_k\right| \\ &\leq \|y\|_\infty\left[\sup_{n\in\mathbb{N}}\sum_k\left|\sum_{j=k}^{\infty}(-1)^{j-k}D_{j-k}T_k a_{nj}\right|\right] < \infty. \end{aligned}$$

Hence, $A \in (c(N^t) : \ell_\infty)$.

This completes the proof. □

Theorem 2.1.11. *$A = (a_{nk}) \in (c(N^t) : c)$ if and only if the condition (2.14) holds, and*

$$F \in (c : c). \tag{2.18}$$

Proof. Suppose that the conditions (2.14) and (2.18) hold, and take any $x = (x_k) \in c(N^t)$. The condition (2.14) implies the existence of A-transform of x.

Therefore, one can derive by using the hypothesis (2.4) with f_{nk} instead of a_{nk} that

$$\sum_{k=0}^{m}|a_k| \leq \sup_{n\in\mathbb{N}}\sum_k |f_{nk}| < \infty$$

for all $m \in \mathbb{N}$. Hence, $(a_k) \in \ell_1$ which implies that $(a_k y_k) \in \ell_1$. Then, we derive by letting $n \to \infty$ on (2.17) with (2.4) and f_{nk} instead of a_{nk} that

$$\lim_{n\to\infty}(Ax)_n = \lim_{n\to\infty}\sum_k f_{nk}y_k = \sum_k a_k y_k. \tag{2.19}$$

Since $(a_k y_k) \in \ell_1$, (2.19) gives that $Ax \in c$, that is, $A \in (c(N^t) : c)$.

Conversely, suppose that $A = (a_{nk}) \in (c(N^t) : c)$ and take $x \in c(N^t)$. Since the inclusion relation $c \subset \ell_\infty$ holds, the necessity of the conditions (2.14) and (2.4) with f_{nk} instead of a_{nk} follows from Theorem 2.1.10.

Now, consider the convergent sequences $u = (u_k) = \left\{u_k^{(n)}\right\}_{k\in\mathbb{N}}$ defined by (2.3) and $x = (x_k) = \left\{\sum_{j=0}^{k} (-1)^{k-j}D_{k-j}T_j\right\}$. Since A-transforms of u and x exist, and belong to the space c by the hypothesis, one can see that $Au = \left\{\sum_{j=k}^{\infty}(-1)^{j-k}D_{j-k}T_k a_{nj}\right\}_{n\in\mathbb{N}} \in c$ and $Ax = \left(\sum_k f_{nk}\right)_{n\in\mathbb{N}} \in c$ which show the necessity of the conditions (2.4) and (2.6) with f_{nk} instead of a_{nk}, respectively. Hence, $F \in (c : c)$.

This completes the proof. □

Corollary 2.1.12. *$A = (a_{nk}) \in (c(N^t) : c_0)$ if and only if (2.14) holds and (2.7) and (2.8) also hold with f_{nk} instead of a_{nk}, respectively.*

Now, we can give the theorem characterizing the class of matrix transformations from a given sequence space μ to the Nörlund space $c(N^t)$.

Theorem 2.1.13. *Suppose that μ be any given sequence space. Then, $A \in (\mu : c(N^t))$ if and only if $G \in (\mu : c)$.*

Proof. Let $x = (x_k) \in \mu$. Consider the following equality

$$\sum_{j=0}^{n}\frac{t_{n-j}}{T_n}\sum_{k=0}^{m}a_{jk}x_k = \sum_{k=0}^{m}g_{nk}x_k \text{ for all } m, n \in \mathbb{N}. \tag{2.20}$$

Then, by letting $m \to \infty$ in (2.20) one can see that $\{N^t(Ax)\}_n = (Gx)_n$ for all $n \in \mathbb{N}$. Therefore, it is immediate that $Ax \in c(N^t)$ whenever $x \in \mu$ if and only if $N^t(Ax) = Gx \in c$ whenever $x \in \mu$.

This completes the proof. □

By combining Theorems 2.1.10, 2.1.11 and Corollary 2.1.12 with Theorem 2.1.13, the following results are derived on the characterization of some matrix classes concerning the space $c(N^t)$ of Nörlund convergent sequences:

Corollary 2.1.14. *Let $A = (a_{nk})$ be an infinite matrix over the complex field. Then, the following statements hold:*

(i) $A \in (c(N^t) : bs)$ *if and only if (2.14) holds and (2.15) also holds with* SF *instead of* F.

(ii) $A \in (c(N^t) : bv_\infty)$ *if and only if (2.14) holds and (2.15) also holds with* ∇F *instead of* F, *where* bv_∞ *denotes the space of all sequences* $x = (x_k)$ *such that* $\nabla x \in \ell_\infty$ (cf. Başar and Altay [25]).

(iii) $A \in (c(N^t) : X_\infty)$ *if and only if (2.14) holds and (2.15) also holds with* CF *instead of* F, *where* X_∞ *denotes the space of all sequences* $x = (x_k)$ *such that* $Cx \in \ell_\infty$ (cf. Ng and Lee [115]).

(iv) $A \in (c(N^t) : r_\infty^q)$ *if and only if (2.14) holds and (2.15) also holds with* R^qF *instead of* F, *where* r_∞^q *denotes the space of all sequences* $x = (x_k)$ *such that* $R^q x \in \ell_\infty$ (cf. Altay and Başar [6]).

(v) $A \in (c(N^t) : a_\infty^r)$ *if and only if (2.14) holds and (2.15) also holds with* A^rF *instead of* F, *where* a_∞^r *denotes the space of all sequences* $x = (x_k)$ *such that* $A^r x \in \ell_\infty$ (cf. Aydın and Başar [16]).

(vi) $A \in (c(N^t) : e_\infty^r)$ *if and only if (2.14) holds and (2.15) also holds with* E^rF *instead of* F, *where* e_∞^r *denotes the space of all sequences* $x = (x_k)$ *such that* $E^r x \in \ell_\infty$ (cf. Altay et al. [11]).

Corollary 2.1.15. *Let $A = (a_{nk})$ be an infinite matrix over the complex field. Then, the following statements hold:*

(i) $A \in (c(N^t) : cs)$ *if and only if (2.14) holds and (2.18) also holds with* SF *instead of* F.

(ii) $A \in (c(N^t) : c(\nabla))$ *if and only if (2.14) holds and (2.18) also holds with* ∇F *instead of* F, *where* $c(\nabla)$ *denotes the space of all sequences* $x = (x_k)$ *such that* $\nabla x \in c$ (cf. Başar [21]).

(iii) $A \in (c(N^t) : \widetilde{c})$ *if and only if (2.14) holds and (2.18) also holds with* CF *instead of* F, *where* $\widetilde{c}$ *denotes the space of all sequences* $x = (x_k)$ *such that* $Cx \in c$ (cf. Şengönül and Başar [128]).

(iv) $A \in (c(N^t) : r_c^q)$ *if and only if (2.14) holds and (2.18) also holds with* R^qF *instead of* F, *where* r_c^q *denotes the space of all sequences* $x = (x_k)$ *such that* $R^q x \in c$ (cf. Altay and Başar [6]).

(v) $A \in (c(N^t) : a_c^r)$ *if and only if (2.14) holds and (2.18) also holds with* A^rF *instead of* F, *where* a_c^r *denotes the space of all sequences* $x = (x_k)$ *such that* $A^r x \in c$ (cf. Aydın and Başar [13]).

(vi) $A \in (c(N^t) : e_c^r)$ *if and only if (2.14) holds and (2.18) also holds with* E^rF *instead of* F, *where* e_c^r *denotes the space of all sequences* $x = (x_k)$ *such that* $E^r x \in c$ (cf. Altay and Başar [5]).

Corollary 2.1.16. *Let $A = (a_{nk})$ be an infinite matrix over the complex field. Then, the following statements hold:*

(i) $A \in (c(N^t) : cs_0)$ *if and only if the condition (2.14) holds and the conditions (2.6), (2.7) and (2.8) also hold with* $(SF)_{nk}$ *instead of* a_{nk} *for all* $k, n \in \mathbb{N}$, *where* cs_0 *denotes the space of all series converging to zero.*

(ii) $A \in (c(N^t) : c_0(\nabla))$ *if and only if the condition (2.14) holds and the conditions (2.6), (2.7) and (2.8) also hold with* $(\nabla F)_{nk}$ *instead of* a_{nk} *for all* $k, n \in \mathbb{N}$, *where* $c_0(\nabla)$ *denotes the space of all sequences* $x = (x_k)$ *such that* $\nabla x \in c_0$ (cf. Başar [21]).

(iii) $A \in (c(N^t) : \widetilde{c}_0)$ *if and only if the condition (2.14) holds and the conditions (2.6), (2.7) and (2.8) also hold with* $(CF)_{nk}$ *instead of* a_{nk} *for all* $k, n \in \mathbb{N}$, *where* $\widetilde{c}_0$ *denotes the space of all sequences* $x = (x_k)$ *such that* $Cx \in c_0$ (cf. Şengönül and Başar [128]).

(iv) $A \in (c(N^t) : r_0^q)$ *if and only if the condition (2.14) holds and the conditions (2.6), (2.7) and (2.8) also hold with* $(R^qF)_{nk}$ *instead of* a_{nk} *for all* $k, n \in \mathbb{N}$; r_0^q *denotes the space of all sequences* $x = (x_k)$ *such that* $R^q x \in c_0$ (cf. Altay and Başar [6]).

(v) $A \in (c(N^t) : a_0^r)$ *if and only if the condition (2.14) holds and the conditions (2.6), (2.7) and (2.8) also hold with* $(A^rF)_{nk}$ *instead of* a_{nk} *for all* $k, n \in \mathbb{N}$, *where* a_0^r *denotes the space of all sequences* $x = (x_k)$ *such that* $A^r x \in c_0$ (cf. Aydın and Başar [13]).

(vi) $A \in (c(N^t) : e_0^r)$ *if and only if the condition (2.14) holds and the conditions (2.6), (2.7) and (2.8) also hold with* $(E^rF)_{nk}$ *instead of* a_{nk} *for all* $k, n \in \mathbb{N}$, *where* e_0^r *denotes the space of all sequences* $x = (x_k)$ *such that* $E^r x \in c_0$ (cf. Altay and Başar [5]).

2.1.4 The Spaces of Nörlund Almost Null and Nörlund Almost Convergent Sequences

Following Tuğ and Başar [131], we introduce the spaces $f_0(N^t)$ and $f(N^t)$ as the set of all N^t-almost null and N^t-almost convergent sequences, respectively, that is,

$$f_0(N^t) := \left\{ (x_k) \in \omega : \lim_{m\to\infty} \sum_{j=0}^{m} \sum_{k=0}^{n+j} \frac{t_{n+j-k}}{(m+1)T_{n+j}} x_k = 0 \text{ uniformly in } n \right\},$$

$$f(N^t) := \left\{ (x_k) \in \omega : \exists l \in \mathbb{C} \ni \lim_{m\to\infty} \sum_{j=0}^{m} \sum_{k=0}^{n+j} \frac{t_{n+j-k}}{(m+1)T_{n+j}} x_k = l \text{ uniformly in } n \right\}.$$

Theorem 2.1.17. *The spaces $f_0(N^t)$ and $f(N^t)$ are BK-spaces with the norm given by*

$$\|x\|_{f(N^t)} = \|N^t x\|_f = \sup_{m,n\in\mathbb{N}} |t_{mn}(N^t x)|,$$

where

$$t_{mn}(N^t x) = \frac{1}{m+1}\sum_{j=0}^{m}\frac{1}{T_{n+j}}\sum_{k=0}^{n+j} t_{n+j-k}x_k \text{ for all } m,n \in \mathbb{N}.$$

Proof. Since f_0 and f are BK-spaces with the norm $\|\cdot\|_\infty$ (see Boos [35, Example 7.3.2 (b)]) and N^t is a triangle matrix, Theorem 4.3.2 of Wilansky [135, p. 61] gives that $f_0(N^t)$ and $f(N^t)$ are BK-spaces with respect to the norm $\|\cdot\|_{f(N^t)}$. □

Let λ denote any of the spaces f_0 or f. With the notation of (2.1), since the transformation $T : \lambda(N^t) \to \lambda$ defined by $x \mapsto y = Tx = N^t x$ is a norm preserving linear bijection, we have the following:

Corollary 2.1.18. *The sequence spaces $f_0(N^t)$ and $f(N^t)$ are linearly norm isomorphic to the spaces f_0 and f, respectively, that is, $f_0(N^t) \cong f_0$ and $f(N^t) \cong f$.*

Now, we can comment on the existence of the Schauder bases of the spaces $f_0(N^t)$ and $f(N^t)$. It is known from Corollary 3.3 of Başar and Kirişçi [29] that the Banach space f has no Schauder basis. Combining this fact and Remark 2.1.4, one can immediately conclude that both of the spaces $f_0(N^t)$ and $f(N^t)$ have no Schauder basis.

Theorem 2.1.19. *The following statements hold:*

(i) The inclusion $f_0(N^t) \subset f(N^t)$ is strict.

(ii) The inclusion $c(N^t) \subset f(N^t)$ is strict.

(iii) The sequence spaces $f(N^t)$ and ℓ_∞ overlap, but neither of them contains the other.

Proof. (i) If we take a sequence $x = (x_k)$ in the space $f_0(N^t)$, then we have $N^t x \in f_0$ and since the inclusion $f_0 \subset f$ trivially holds, then we have $N^t x \in f$ which gives us $x \in f(N^t)$. Thus, the inclusion $f_0(N^t) \subset f(N^t)$ holds.

Now, we consider a sequence $x = (x_k)$ in the space $f(N^t)$ but is not in the space $f_0(N^t)$, that is, we show that $f(N^t) \setminus f_0(N^t)$ is not empty. Consider the sequence $x = e$. Then, $N^t e = e \in f \setminus f_0$, so we have $x \in f(N^t) \setminus f_0(N^t)$. Hence, the inclusion $f_0(N^t) \subset f(N^t)$ strictly holds.

(ii) If $x \in c(N^t)$, then we have $N^t x \in c$ and the inclusion $c \subset f$ is well known, so that $N^t x \in f$, i.e., $x \in f(N^t)$. Hence, the inclusion $c(N^t) \subset f(N^t)$ holds.

Now, we should show that the set $f(N^t)\backslash c(N^t)$ is not empty. For this, we consider the sequence $x = (x_k)$ defined by

$$x_k = \sum_{j=0}^{k}(-1)^k D_{k-j}T_j \quad \textit{for all} \quad k \in \mathbb{N}. \tag{2.21}$$

Therefore, we obtain that

$$(N^t x)_n = \sum_{k=0}^{n}\frac{t_{n-k}}{T_n}\sum_{j=0}^{k}(-1)^k D_{k-j}T_j = (-1)^n \quad \textit{for all} \quad n \in \mathbb{N}.$$

Then, it is clear that $N^t x$ is not in c but

$$\begin{aligned}
\lim_{m\to\infty}\sum_{k=0}^{m}\sum_{j=0}^{k+n}\frac{t_{k+n-j}}{(m+1)T_{k+n}}\sum_{j=0}^{k}(-1)^k D_{k-j}T_j &= \lim_{m\to\infty}\sum_{k=0}^{m}\frac{(-1)^{n+k}}{m+1} \\
&= \lim_{m\to\infty}\frac{(-1)^n\left[1+(-1)^m\right]}{2(m+1)} \\
&= 0 \text{ uniformly in } n.
\end{aligned}$$

That is to say that $N^t x \in f$. Thus, the set $f(N^t)\backslash c(N^t)$ is not empty.

(iii) To prove this, first we prove that $f(N^t)$ and ℓ_∞ are not disjoint. If we take $z = e$ then, since $N^t z = e \in f$, $z \in f(N^t)$. Furthermore, it is trivial that $z \in \ell_\infty$. This shows that there exists at least one point belonging to both $f(N^t)$ and ℓ_∞, as was asserted.

Now, we prove that the sequence spaces $f(N^t)$ and ℓ_∞ do not include each other. Let us consider the sequence $x = (x_k)$ defined by (2.21). Then, since $N^t x = \{(-1)^n\} \in f$, $x \in f(N^t)$ but $x \notin \ell_\infty$. Hence, $x \in f(N^t)\backslash\ell_\infty$.

Now, we consider the sequence $s = U^t v$ with

$$v = (0, 0, \ldots, 0, 1, 1, \ldots, 1, 0, 0, \ldots, 0, 1, 1, \ldots, 1, 0, 0, \ldots),$$

where the sequence v is defined by Miller and Orhan [102], and belongs to the set $\ell_\infty \setminus f$ and the blocks of 0's are increasing by factors of 100 and the blocks of 1's are increasing by factors of 10. Then, it is clear that $s \in \ell_\infty$ but is not in the space $f(N^t)$. This shows that the spaces $f(N^t)$ and ℓ_∞ do not include each other.

This completes the proof. □

2.1.5 The Alpha-, Beta- and Gamma-duals of the Spaces $f_0(N^t)$ and $f(N^t)$

In the present subsection, we determine the alpha-, beta- and gamma-duals of the spaces $f_0(N^t)$ and $f(N^t)$. We start with quoting the following two lemmas whose some parts related with the characterization of matrix transformations on/in the space f.

Lemma 2.1.20. *Let $A = (a_{nk})$ be an infinite matrix over the complex field. Then, the following statements hold:*

(i) $A \in (f_0 : \ell_1) = (f : \ell_1)$ *if and only if*

$$\sup_{K \in \mathcal{F}} \sum_n \left| \sum_{k \in K} a_{nk} \right| < \infty.$$

(ii) $A \in (\ell_\infty : \ell_\infty) = (f : \ell_\infty)$ *if and only if (2.4) holds.*

(iii) (cf. Sıddıqi [121]) $A \in (f : c) = (f_0 : c)$ *if and only if (2.4), (2.5), (2.6) hold and*

$$\lim_{n \to \infty} \sum_k |\Delta(a_{nk} - \alpha_k)| = 0. \tag{2.22}$$

Here and after, Δ *denotes the forward difference matrix, i.e.,* $\Delta(a_{nk} - \alpha_k) = a_{n,k+1} - \alpha_{k+1} - (a_{nk} - \alpha_k)$ *for all* $n, k \in \mathbb{N}$.

(iv) $A \in (\ell_\infty : c)$ *if and only if (2.5) holds, and*

$$\lim_{n \to \infty} \sum_k |a_{nk} - \alpha_k| = 0 \text{ uniformly in } n. \tag{2.23}$$

Lemma 2.1.21. *Let $A = (a_{nk})$ be an infinite matrix over the complex field. Then, the following statements hold:*

(i) (cf. Duran [48]) $A \in (\ell_\infty : f)$ *if and only if (2.4) holds and*

$$\exists \alpha_k \in \mathbb{C} \text{ such that } f - \lim a_{nk} = \alpha_k \text{ for each } k \in \mathbb{N}, \tag{2.24}$$

$$\lim_{n \to \infty} \sum_k \left| \frac{1}{m+1} \sum_{j=0}^{m} a_{n+j,k} - \alpha_k \right| = 0 \tag{2.25}$$

uniformly in n.

(ii) (cf. Duran [48]) $A \in (f : f)$ *if and only if (2.4) and (2.24) hold, and*

$$\exists \alpha \in \mathbb{C} \text{ such that } f - \lim \sum_k a_{nk} = \alpha, \tag{2.26}$$

$$\lim_{m \to \infty} \sum_k \left| \Delta \left(\frac{1}{m+1} \sum_{j=0}^{m} a_{n+j,k} - \alpha_k \right) \right| = 0 \tag{2.27}$$

uniformly in n.

(iii) (cf. King [71]) $A \in (c : f)$ *if and only if (2.4), (2.24) and (2.26) hold.*

(iv) (cf. Nanda [110]) $A \in (\ell_p : f)$ *if and only if (2.24) holds, and*

$$\sup_{n\in\mathbb{N}} \sum_k |a_{nk}|^q < \infty, \quad (1 \le p < \infty), \tag{2.28}$$

$$\sup_{k,n\in\mathbb{N}} |a_{nk}| < \infty, \quad (0 < p < 1). \tag{2.29}$$

Theorem 2.1.22. *The α-dual of the spaces $f_0(N^t)$ and $f(N^t)$ is the set*

$$d_1^t := \left\{ a = (a_k) \in \omega : \sup_{K\in\mathcal{F}} \sum_n \left| \sum_{k\in K} (-1)^{n-k} D_{n-k} T_k a_n \right| < \infty \right\}.$$

Proof. Let us define the matrix $B = (b_{nk}^t)$ via $a = (a_k) \in \omega$ by (2.9). Then, since the relation (2.2) holds, it is immediate that (2.10) holds. Therefore, we read by the relation (2.10) that $ax = (a_n x_n) \in \ell_1$ whenever $x = (x_n) \in f_0(N^t)$ or $\in f(N^t)$ if and only if $By \in \ell_1$ whenever $y = (y_k) \in f_0$ or $\in f$. This leads to the fact that $a \in [f(N^t)]^\alpha$ if and only if $B \in (f_0 : \ell_1) = (f : \ell_1)$. Therefore, we derive by Part (i) of Lemma 2.1.20 that

$$\sup_{K\in\mathcal{F}} \sum_n \left| \sum_{k\in K} (-1)^{n-k} D_{n-k} T_k a_n \right| < \infty.$$

This means that the α-dual of the spaces $f_0(N^t)$ and $f(N^t)$ is the set d_1^t, as desired. □

Theorem 2.1.23. *Define the set d_2^t by*

$$d_2^t := \left\{ a = (a_k) \in \omega : \sup_{n\in\mathbb{N}} \sum_k \left| \sum_{j=k}^{n} (-1)^{j-k} D_{j-k} T_k a_j \right| < \infty \right\}.$$

Then, $[f_0(N^t)]^\beta = [f(N^t)]^\beta = d_2^t \cap cs$.

Proof. Let $x = (x_k)$ be any sequence in the space $f_0(N^t)$ or $f(N^t)$. Then, one can immediately observe that $a = (a_k) \in [f_0(N^t)]^\beta = [f(N^t)]^\beta$ if and only if the matrix $B = (b_{nk}^t)$ is in the class $(f : cs)$ of infinite matrices, where b_{nk}^t is defined by (2.9). This is equivalent to the fact that $a = (a_k) \in [f_0(N^t)]^\beta = [f(N^t)]^\beta$ if and only if the matrix $E = (e_{nk}) = SB$ is in the class $(f_0 : c) = (f : c)$ of infinite matrices, where e_{nk} is defined by (2.12). In this situation, we derive from (2.4) and (2.5) with e_{nk}^t instead of a_{nk} that the following conditions are satisfied:

$$\sup_{n\in\mathbb{N}} \sum_k \left| \sum_{j=k}^{n} (-1)^{j-k} D_{j-k} T_k a_j \right| < \infty,$$

$$(a_k) \in cs.$$

This step leads us to the desired consequence that the β-dual of the spaces $f_0(N^t)$ and $f(N^t)$ is the set $d_2^t \cap cs$. □

Theorem 2.1.24. *The gamma-dual of the spaces $f_0(N^t)$ and $f(N^t)$ is the set d_2^t.*

Proof. This is similar to the proof of Theorem 2.1.23 with Part (ii) of Lemma 2.1.20 instead of Part (iii) of Lemma 2.1.20. By avoiding the repetition of the similar statements, we omit details. □

2.1.6 Matrix Transformations Related to the Space $f(N^t)$

Let λ denote any of the classical sequence spaces ℓ_∞, c, c_0 and ℓ_p or any of the sequence spaces f_0 and f. Then, the domain λ_{N^t} is called the *Nörlund sequence space*. Therefore, since $\lambda_{N^t} \cong \lambda$ it is trivial that the two sided implication "$x \in \lambda_{N^t}$ if and only if $y \in \lambda$" holds.

For the sake of brevity in notation, we shall also write here and after that

$$d_{nk} = \sum_{j=0}^{n} \frac{t_{n-j}}{T_n} b_{jk} \quad \text{and} \quad b(n,k) = \sum_{j=0}^{n} b_{jk} \tag{2.30}$$

for all $k, n \in \mathbb{N}$.

Following Yeşilkayagil and Başar [136], we shall employ the concept of the pair of summability matrices (PSM) defined by a relation between two infinite matrices such that one of them is applied to the sequences in a Nörlund space and the other one applied to the sequences in a space which is isomorphic to the Nörlund space. We also give a basic theorem related to the PSM. Therefore, we characterize the classes $(\lambda(N^t) : \mu)$ and $(\mu : \lambda(N^t))$ of infinite matrices. In the present subsection, we suppose that λ and μ are given two sequence spaces.

Now, we may focus on the PSM. Let us suppose that the infinite matrices $A = (a_{nk})$ and $B = (b_{nk})$ transform the sequences $x = (x_k)$ and $y = (y_k)$ which be connected with the relation (2.1) to the sequences $u = (u_n)$ and $v = (v_n)$, respectively, i.e.,

$$u_n = \big(Ax\big)_n = \sum_k a_{nk} x_k \quad \text{for each} \quad n \in \mathbb{N}, \tag{2.31}$$

$$v_n = \big(By\big)_n = \sum_k b_{nk} y_k \quad \text{for each} \quad n \in \mathbb{N}. \tag{2.32}$$

It is clear here that the method B is applied to the N^t-transform of the sequence $x = (x_k)$ while the method A is directly applied to the terms of the sequence $x = (x_k)$.

Let us assume that the matrix product BN^t exists. We say in this situation that the matrices A and B in (2.31), (2.32) are the *PSM* if u_n is reduced to v_n (or v_n is reduced to u_n) under the application of the formal summation by parts. This leads us to the fact that BN^t exists and is equal to A and $(BN^t)x = B(N^t x)$ formally holds, if one side exists. Therefore, we have the relation

$$a_{nk} = \sum_{j=k}^{\infty} \frac{t_{j-k}}{T_j} b_{nj} \text{ or } b_{nk} = \sum_{j=k}^{\infty} (-1)^{j-k} D_{j-k} T_k a_{nj} \tag{2.33}$$

for all $k, n \in \mathbb{N}$. By taking into account the relation (2.1) one can derive that

$$\sum_{k=0}^{m} b_{nk} y_k = \sum_{k=0}^{m} b_{nk} \left(\frac{1}{T_k} \sum_{j=0}^{k} t_{k-j} x_j \right) = \sum_{k=0}^{m} \sum_{j=k}^{m} \frac{t_{j-k}}{T_j} b_{nj} x_k \tag{2.34}$$

for all $m, n \in \mathbb{N}$. Therefore, we obtain by (2.34) as $m \to \infty$ that v_n reduces to u_n, as follows:

$$v_n = \sum_k b_{nk} y_k = \sum_k b_{nk} \left(\frac{1}{T_k} \sum_{j=0}^{k} t_{k-j} x_j \right) = \sum_k \sum_{j=k}^{\infty} \frac{t_{j-k}}{T_j} b_{nj} x_k = u_n.$$

But, the order of summation may not be reversed. So, the methods A and B are not necessarily equivalent.

Theorem 2.1.25. *Let the elements of the matrices $A = (a_{nk})$ and $B = (b_{nk})$ be connected with the relation (2.33). Then, $A \in (\lambda(N^t) : \mu)$ if and only if $B \in (\lambda : \mu)$.*

Proof. Let $A = (a_{nk})$ and $B = (b_{nk})$ be a PSM.

Suppose that $A \in (\lambda(N^t) : \mu)$. Then, Ax exists and belongs to μ for all $x \in \lambda(N^t)$. Therefore, we again have the equality (2.16) derived from m^{th} partial sum of the series $\sum_k a_{nk} x_k$ with the relation (2.2). Then, we have from (2.16) by letting $m \to \infty$ that $Ax = By$. Therefore, it is immediate that $By \in \mu$ whenever $y \in \lambda$, i.e., $B \in (\lambda : \mu)$.

Conversely, suppose that $B \in (\lambda : \mu)$. Then, By exists and belongs to μ for all $y \in \lambda$. Therefore, by letting $m \to \infty$ in (2.34), we get $By = Ax$ which gives the desired fact that $A \in (\lambda(N^t) : \mu)$. □

By interchanging the spaces λ_{N^t} and λ with the space μ, we have:

Theorem 2.1.26. *Suppose that the elements of the infinite matrices $A = (a_{nk})$ and $C = (c_{nk})$ are connected with the relation*

$$c_{nk} = \sum_{j=0}^{n} \frac{t_{n-j}}{T_n} a_{jk} \quad \textit{for all} \quad k, n \in \mathbb{N}. \tag{2.35}$$

Then, $A \in (\mu : \lambda(N^t))$ if and only if $C \in (\mu : \lambda)$.

Proof. Let us take any $s = (s_k) \in \mu$ and consider the following equality with (2.35) that

$$\sum_{k=0}^{m} c_{nk} s_k = \sum_{j=0}^{n} \sum_{k=0}^{m} \frac{t_{n-j}}{T_n} a_{jk} s_k \quad \text{for all} \quad m, n \in \mathbb{N}$$

which yields as $m \to \infty$ that $(Cs)_n = \{N^t(As)\}_n$ for all $n \in \mathbb{N}$. Now, we immediately deduce from here that $As \in \lambda(N^t)$ whenever $s \in \mu$ if and only if $Cs \in \lambda$ whenever $s \in \mu$.

This step completes the proof. □

Of course, Theorems 2.1.25 and 2.1.26 have several consequences depending on the choice of the sequence spaces λ and μ. By Theorem 2.1.25, the necessary and sufficient conditions for $A \in (\lambda(N^t) : \mu)$ may be derived by replacing the elements of A by those of the elements of $B = AU^t$, where the necessary and sufficient conditions on the matrix B are read from the concerning results in the existing literature. Since Theorems 2.1.25 and 2.1.26 are respectively related with the matrix transformations on the Nörlund sequence spaces and into itself, the characterizations of the matrix mappings between the Nörlund sequence spaces may be derived by combining Theorems 2.1.25 and 2.1.26. Now, we may quote the results on the characterization of some matrix classes concerning the Nörlund sequence spaces.

The rest of this subsection, we assume that the elements of the matrices $A = (a_{nk})$, $B = (b_{nk})$, $C = (c_{nk})$ and $D = (d_{nk})$ are connected with the relations (2.30), (2.33) and (2.35); and $q = p/(p-1)$ when $1 \leq p < \infty$.

Corollary 2.1.27. *Let $A = (a_{nk})$ be an infinite matrix over the complex field. Then, the following statements hold:*

(i) $A \in (f(N^t) : \ell_\infty)$ *if and only if (2.4) holds with* b_{nk} *instead of* a_{nk}.

(ii) $A \in (f(N^t) : f)$ *if and only if (2.4), (2.24), (2.26) and (2.27) hold with* b_{nk} *instead of* a_{nk}.

(iii) $A \in (f(N^t) : f; p)$ *if and only if (2.4), (2.24), (2.26) and (2.27) hold with* b_{nk} *instead of* a_{nk}*, and* $\alpha_k = 0$ *for all* $k \in \mathbb{N}$, $\alpha = 1$.

(iv) $A \in (f(N^t) : c)$ *if and only if (2.4)-(2.6) and (2.22) hold with* b_{nk} *instead of* a_{nk}.

(v) $A \in (f(N^t) : c; p)$ *if and only if (2.4)-(2.6) and (2.22) hold with* b_{nk} *instead of* a_{nk}*, and* $\alpha_k = 0$ *for all* $k \in \mathbb{N}$, $\alpha = 1$.

(vi) $A \in (f(N^t) : c_0)$ *if and only if (2.4)-(2.6) and (2.22) hold with* b_{nk} *instead of* a_{nk}*, and* $\alpha_k = 0$ *for all* $k \in \mathbb{N}$, $\alpha = 0$.

(vii) $A \in (\ell_\infty(N^t) : f)$ *if and only if (2.4), (2.24) and (2.25) hold with* b_{nk} *instead of* a_{nk}.

(viii) $A \in (\ell_\infty(N^t) : c)$ *if and only if (2.5) and (2.23) hold with* b_{nk} *instead of* a_{nk}.

(ix) $A \in (\ell_p(N^t) : f)$ *if and only if (2.24), (2.28) and (2.29) hold with* b_{nk} *instead of* a_{nk}.

Corollary 2.1.28. *Let $A = (a_{nk})$ be an infinite matrix over the complex field. Then, the following statements hold:*

(i) $A \in (f(N^t) : \ell_\infty(N^t))$ *if and only if (2.4) holds with* d_{nk} *instead of* a_{nk}.

(ii) $A \in (f(N^t) : f(N^t))$ *if and only if (2.4), (2.24), (2.26) and (2.27) hold with* d_{nk} *instead of* a_{nk}.

(iii) $A \in (f(N^t) : f(N^t); p)$ *if and only if (2.4), (2.24), (2.26) and (2.27) hold with* d_{nk} *instead of* a_{nk}, *and* $\alpha_k = 0$ *for all* $k \in \mathbb{N}$, $\alpha = 1$.

(iv) $A \in (f(N^t) : c(N^t))$ *if and only if (2.4)-(2.6) and (2.22) hold with* d_{nk} *instead of* a_{nk}.

(v) $A \in (f(N^t) : c(N^t); p)$ *if and only if (2.4)-(2.6) and (2.22) hold with* d_{nk} *instead of* a_{nk}, *and* $\alpha_k = 0$ *for all* $k \in \mathbb{N}$, $\alpha = 1$.

(vi) $A \in (f(N^t) : c_0(N^t))$ *if and only if (2.4)-(2.6) and (2.22) hold with* d_{nk} *instead of* a_{nk}, *and* $\alpha_k = 0$ *for all* $k \in \mathbb{N}$, $\alpha = 0$.

(vii) $A \in (\ell_\infty(N^t) : f(N^t))$ *if and only if (2.4), (2.24) and (2.25) hold with* d_{nk} *instead of* a_{nk}.

(viii) $A \in (\ell_p(N^t) : f(N^t))$ *if and only if (2.24), (2.28) and (2.29) hold with* d_{nk} *instead of* a_{nk}, *respectively.*

Corollary 2.1.29. *Let $A = (a_{nk})$ be an infinite matrix over the complex field. Then, the following statements hold:*

(i) $A \in (f(N^t) : bs)$ *if and only if (2.4) holds with* $b(n,k)$ *instead of* a_{nk}.

(ii) $A \in (f(N^t) : fs)$ *if and only if (2.4), (2.24), (2.26) and (2.27) hold with* $b(n,k)$ *instead of* a_{nk}.

(iii) $A \in (f(N^t) : fs; p)$ *if and only if (2.4), (2.24), (2.26) and (2.27) hold with* $b(n,k)$ *instead of* a_{nk}, *and* $\alpha_k = 0$ *for all* $k \in \mathbb{N}$, $\alpha = 1$.

(iv) $A \in (f(N^t) : cs)$ *if and only if (2.4)-(2.6) and (2.22) hold with* $b(n,k)$ *instead of* a_{nk}.

(v) $A \in (f(N^t) : cs; p)$ *if and only if (2.4)-(2.6) and (2.22) hold with* $b(n,k)$ *instead of* a_{nk}, *and* $\alpha_k = 0$ *for all* $k \in \mathbb{N}$, $\alpha = 1$.

(vi) $A \in (f(N^t) : cs_0)$ *if and only if (2.4)-(2.6) and (2.22) hold with* $b(n,k)$ *instead of* a_{nk}, *and* $\alpha_k = 0$ *for all* $k \in \mathbb{N}$, $\alpha = 0$.

(vii) $A \in (\ell_\infty(N^t) : fs)$ *if and only if (2.4), (2.24) and (2.25) hold with* $b(n,k)$ *instead of* a_{nk}.

(viii) $A \in (\ell_p(N^t) : fs)$ *if and only if (2.24), (2.28) and (2.29) hold with* $b(n,k)$ *instead of* a_{nk}.

Corollary 2.1.30. *Let $A = (a_{nk})$ be an infinite matrix over the complex field. Then, the following statements hold:*

(i) $A \in (\ell_\infty : f(N^t))$ *if and only if (2.4), (2.24) and (2.25) hold with* c_{nk} *instead of* a_{nk}.

(ii) $A \in (f : f(N^t))$ *if and only if (2.4), (2.24), (2.26) and (2.27) hold with* c_{nk} *instead of* a_{nk}.

(iii) $A \in (f : f(N^t); p)$ *if and only if (2.4), (2.24), (2.26) and (2.27) hold with* c_{nk} *instead of* a_{nk}, *and* $\alpha_k = 0$ *for all* $k \in \mathbb{N}$, $\alpha = 1$.

(iv) $A \in (c : f(N^t)) = (c_0 : f(N^t))$ *if and only if (2.4), (2.24) and (2.26) hold with* c_{nk} *instead of* a_{nk}.

(v) $A \in (c : f(N^t); p)$ *if and only if (2.4), (2.24) and (2.26) hold with* c_{nk} *instead of* a_{nk}, *and* $\alpha_k = 0$ *for all* $k \in \mathbb{N}$, $\alpha = 1$.

(vi) $A \in (f : \ell_\infty(N^t))$ *if and only if (2.4) holds with* c_{nk} *instead of* a_{nk}.

Finally, we mention Steinhaus type theorems which were formulated by Maddox [83], as follows: Let λ and μ be any two sequence spaces having some notion of limit or sum, and $(\lambda : \mu; p)$ denotes the class of regular matrices and ν also be any sequence space such that $\nu \supset \lambda$. Then, a result of the form $(\lambda : \mu; p) \cap (\nu : \mu) = \emptyset$ is called a *theorem of Steinhaus type.*

Now, we can give the following theorem including two Steinhaus type conclusions.

Theorem 2.1.31. *The following statements hold:*

(i) *The classes* $(f(N^t) : c; p)$ *and* $(\ell_\infty(N^t) : c)$ *are disjoint.*

(ii) *The classes* $(f(N^t) : f; p)$ *and* $(\ell_\infty(N^t) : f)$ *are disjoint.*

Proof. (i) Suppose conversely that the classes $(f(N^t) : c; p)$ and $(\ell_\infty(N^t) : c)$ are not disjoint. Then, there exists at least one infinite matrix A satisfying the conditions of Parts (v) and (viii) of Corollary 2.1.27. Therefore, we derive by using the condition (2.23) with $\alpha_k = 0$ and with f_{nk} instead of a_{nk} that

$$\lim_{n\to\infty} \sum_{k=0}^{\infty} |f_{nk}| = 0. \tag{2.36}$$

Nevertheless, from Part (v) of Corollary 2.1.27 with $\alpha = 1$ and with f_{nk} instead of a_{nk} we have

$$\lim_{n\to\infty} \sum_k f_{nk} = 1$$

which contradicts (2.36).

This completes the proof of Part (i).

(ii) This is similar to the proof of Part (i) of the present theorem with Parts (iii) and (vii) instead of Parts (v) and (viii) of Corollary 2.1.27, respectively. So, we omit details. □

2.2 Domains of the Euler-Cesàro Difference Matrix in the Classical Sequence Spaces

In this section, we put emphasis on the spaces of Euler-Cesàro difference null, convergent, bounded and absolutely p-summable sequences.

The inverses $E_1^{-1} = (g_{nk})$ and $C^{-1} = (h_{nk})$ of first order Euler matrix E_1 and Cesàro matrix C are given by

$$g_{nk} = \begin{cases} \binom{n}{k}(-1)^{n-k}2^k & , \quad 0 \leq k \leq n, \\ \\ 0 & , \quad k > n, \end{cases}$$
$$h_{nk} = \begin{cases} (-1)^{n-k}(k+1) & , \quad n-1 \leq k \leq n, \\ \\ 0 & , \quad 0 \leq k \leq n-2 \text{ or } k > n \end{cases}$$

for all $k, n \in \mathbb{N}$. We define the matrix $\widetilde{B} = (\widetilde{b}_{nk})$ by the composition of the matrices E_1, C and ∇ as

$$\widetilde{b}_{nk} = \begin{cases} \dfrac{\binom{n}{k}}{2^n(k+1)} & , \quad 0 \leq k \leq n, \\ \\ 0 & , \quad k > n \end{cases} \tag{2.1}$$

for all $k, n \in \mathbb{N}$. Define the sequence $y = (y_n)$ by the $\widetilde{B}$-transform of a sequence $x = (x_k)$, i.e.,

$$y_n = (\widetilde{B}x)_n = \sum_{k=0}^{n} \frac{\binom{n}{k}}{2^n(k+1)} x_k \quad \text{for each} \quad n \in \mathbb{N}. \tag{2.2}$$

We suppose throughout in this section that the sequences $x = (x_k)$ and $y = (y_k)$ are connected with the relation (2.2). One can obtain by a straightforward calculation from (2.2) that

$$x_n = (\widetilde{B}^{-1}y)_n = \sum_{k=0}^{n}\sum_{j=k}^{n} \binom{j}{k}(-1)^{j-k}2^k(2j-k+1)y_k \tag{2.3}$$

for each $n \in \mathbb{N}$. The relation (2.3) gives that the inverse $\widetilde{B}^{-1} = (d_{nk})$ of $\widetilde{B}$ in (2.1) is defined by

$$d_{nk} = \begin{cases} \sum_{j=k}^{n} \binom{j}{k}(-1)^{j-k}2^k(2j-k+1) & , \quad 0 \leq k \leq n, \\ \\ 0 & , \quad k > n \end{cases} \tag{2.4}$$

for all $k, n \in \mathbb{N}$.

2.2.1 The Euler-Cesàro Difference Spaces of Null, Convergent and Bounded Sequences

In this subsection, following Başar and Braha [27], we define the spaces $\breve{c}_0$, $\breve{c}$ and $\breve{\ell}_\infty$ of Euler-Cesàro null, Euler-Cesàro convergent and Euler-Cesàro bounded difference sequences. Also, we present some inclusion theorems and construct the Schauder bases of the spaces $\breve{c}_0$ and $\breve{c}$.

Now, we introduce the spaces $\breve{c}_0$, $\breve{c}$ and $\breve{\ell}_\infty$ of Euler-Cesàro null, Euler-Cesàro convergent and Euler-Cesàro bounded difference sequences as the set of all sequences whose $\widetilde{B}$-transforms are in the spaces c_0, c and ℓ_∞, respectively, i.e.,

$$\begin{aligned} \breve{c}_0 &:= \left\{ x = (x_n) \in \omega : \lim_{n\to\infty} \sum_{k=0}^{n} \frac{\binom{n}{k}}{2^n(k+1)} x_k = 0 \right\}, \\ \breve{c} &:= \left\{ x = (x_n) \in \omega : \exists l \in \mathbb{C} \text{ such that } \lim_{n\to\infty} \sum_{k=0}^{n} \frac{\binom{n}{k}}{2^n(k+1)} x_k = l \right\}, \\ \breve{\ell}_\infty &:= \left\{ x = (x_n) \in \omega : \sup_{n\in\mathbb{N}} \left| \sum_{k=0}^{n} \frac{\binom{n}{k}}{2^n(k+1)} x_k \right| < \infty \right\}. \end{aligned}$$

With the notation of (1.3), the spaces $\breve{c}_0$, $\breve{c}$ and $\breve{\ell}_\infty$ can be redefined as follows:

$$\breve{c}_0 = (c_0)_{\widetilde{B}}, \quad \breve{c} = c_{\widetilde{B}} \text{ and } \breve{\ell}_\infty = (\ell_\infty)_{\widetilde{B}}.$$

Throughout Subsection 2.2.1, by λ and μ we denote any of the spaces $\breve{c}_0$, $\breve{c}$ or $\breve{\ell}_\infty$ and c_0, c or ℓ_∞, respectively.

Theorem 2.2.1. *The set λ is a linear space with the coordinatewise addition and scalar multiplication which is the BK-space with the norm $\|x\|_\lambda = \|\widetilde{B}x\|_\infty$.*

Proof. Proof of the first part of the theorem is a routine verification and so we omit details. Since c_0, c and ℓ_∞ are the BK-spaces with respect to their usual sup-norm and $\widetilde{B}$ is a triangle matrix, Theorem 4.3.2 of Wilansky [135, p. 61] gives the fact that $\breve{c}_0$, $\breve{c}$ and $\breve{\ell}_\infty$ are the BK-spaces.

This completes the proof. □

Remark 2.2.2. *One can easily check that the absolute property does not hold on the space λ, that is $\|x\|_\lambda \neq \||x|\|_\lambda$ for at least one sequence in the space λ, and this says that λ is a sequence space of non-absolute type, where $|x| = (|x_k|)$.*

With the notation of (2.2), since the transformation T defined from λ to μ by $x \mapsto y = Tx = \widetilde{B}x$ is a linear bijection, we have the following:

Corollary 2.2.3. *The sequence space λ is linearly norm isomorphic to the space μ.*

Now, we give some inclusion relations concerning the spaces λ and μ.

Theorem 2.2.4. *The inclusion $\mu \subset \lambda$ strictly holds.*

Proof. Let $x = (x_k) \in \mu$. Then, since it is immediate that

$$\begin{aligned} \|x\|_\lambda = \|\widetilde{B}x\|_\infty &= \sup_{n\in\mathbb{N}} \left| \sum_{k=0}^{n} \frac{\binom{n}{k}}{2^n(k+1)} x_k \right| \\ &\leq \|x\|_\infty \sup_{n\in\mathbb{N}} \sum_{k=0}^{n} \frac{\binom{n}{k}}{2^n(k+1)} \\ &= \|x\|_\infty \sup_{n\in\mathbb{N}} \frac{2^{n+1}-1}{2^n(n+1)} < \infty, \end{aligned}$$

the inclusion $\mu \subset \lambda$ holds. Now, consider the sequences $u = (u_k) = \{[1 + (-1)^{k+1}]/2\}_{k\in\mathbb{N}}$ and $v = (v_k) = (k+1)_{k\in\mathbb{N}}$. Thus, we obtain for all $n \in \mathbb{N}$ that

$$\begin{aligned} (\widetilde{B}u)_n &= \frac{1}{2^n} \sum_{k=0}^{n} \frac{\binom{n}{k}[1+(-1)^{k+1}]}{2(k+1)} \\ &= \frac{1}{2^{n+1}} \left[\sum_{k=0}^{n} \frac{\binom{n}{k}}{k+1} + \sum_{k=0}^{n} \frac{\binom{n}{k}(-1)^{k+1}}{k+1} \right] \\ &= \frac{1}{2^{n+1}} \left(\frac{2^{n+1}-1}{n+1} + \frac{n}{n+1} - 1 \right) \\ (\widetilde{B}v)_n &= \frac{1}{2^n} \sum_{k=0}^{n} \frac{\binom{n}{k}(k+1)}{k+1} = 1. \end{aligned}$$

Therefore, since $\widetilde{B}u \in c_0$, $u \in \breve{c}_0$ while $u \notin c_0$ and $\widetilde{B}v \in c \subset \ell_\infty$, v is both in $\breve{c}$ and hence in $\breve{\ell}_\infty$ but v does not neither in c nor in ℓ_∞. This shows that the inclusions $c_0 \subset \breve{c}_0$, $c \subset \breve{c}$ and $\ell_\infty \subset \breve{\ell}_\infty$ are strict. □

As an immediate consequence of Remark 2.1.4, we have the following:

Corollary 2.2.5. *Let $\alpha_n = (\widetilde{B}x)_n$ for all $n \in \mathbb{N}$. Define the sequence $u^{(n)} = \left\{u_k^{(n)}\right\}_{k\in\mathbb{N}}$ in the space $\breve{c}_0$ by*

$$u_k^{(n)} = \begin{cases} \sum_{j=k}^{n} \binom{j}{k}(-1)^{j-k}2^k(2j-k+1) &, \quad 0 \leq k \leq n, \\ 0 &, \quad k > n \end{cases} \tag{2.5}$$

for every fixed $n \in \mathbb{N}$. Then, the following statements hold:

(a) *The sequence $\{u^{(n)}\}_{n\in\mathbb{N}}$ is a basis for the space $\breve{c}_0$ and any $x \in \breve{c}_0$ has a unique representation of the form $x = \sum_n \alpha_n u^{(n)}$.*

(b) *The set* $\{e, u^{(n)}\}$ *is a basis for the space* $\breve{c}$ *and any* $x \in \breve{c}$ *has a unique representation of the form* $x = le + \sum_n (\alpha_n - l)u^{(n)}$, *where* $l = \lim_{k\to\infty} (\widetilde{B}x)_k$.

Since the sequence spaces $\breve{c}_0$ and $\breve{c}$ have Schauder bases, Corollary 2.2.5 directly gives the following:

Corollary 2.2.6. *The sequence spaces* $\breve{c}_0$ *and* $\breve{c}$ *are separable.*

Theorem 2.2.7. *The sequence space* $\breve{c}$ *has not the AK-property.*

Proof. Let $x = (1, 2, 3, 4, \ldots, n, \ldots) \in \breve{c}$. Then, the nth section $x^{[n]}$ of x is $x^{[n]} = (1, 2, \ldots, n, 0, 0, \ldots)$. Therefore, we have

$$\left\|x - x^{[n]}\right\|_{\breve{c}} = \sup_{k\geq n} \frac{1}{2^k} \left| \sum_{j=0}^{k} \binom{k}{j} \frac{1}{j+1} [(j+1) - 0] \right| = \sup_{k\geq n} \frac{1}{2^k} \cdot 2^k = 1.$$

□

In the sequel, we mention some concepts (see [41]) related to the sequence spaces.

A sequence space λ is said to be symmetric if $x_{\sigma(n)} \in \lambda$ whenever $x \in \nu$, where σ is any permutation in $\mathbb{N}$. A *sequence algebra* ν (see Maddox [88, p. 153]) is a linear space together with an internal operation of multiplication of elements of ν such that

$$\begin{aligned} &xy \in \nu, \quad x(yz) = (xy)z, \quad x(y+z) = xy + xz, \quad (x+y)z = xz + yz, \\ &\alpha(xy) = x(\alpha y) \quad \text{for all scalars} \quad \alpha. \end{aligned}$$

Now, we show some of the above properties for the sequence space $\breve{c}$.

Theorem 2.2.8. *The sequence space* $\breve{c}$ *is not monotone nor sequence algebra.*

Proof. To prove the first part, we consider the sequences $x = (x_n) = (n+1)_{n\in\mathbb{N}} \in \breve{c}$ and $y = (y_n) = \{(n+1)^2\}_{n\in\mathbb{N}}$. After some calculations it follows that $y \notin \breve{c}$ which proves that $\breve{c}$ is not monotone. To prove that $\breve{c}$ is not convergence-free we define the sequences: $x = (x_n) = (n+1) = (y_n) = y$. Then, $x = y \in \breve{c}$, but $xy = (1^2, 2^2, 3^2, \ldots) \notin \breve{c}$. Really after some calculations, we get that

$$\begin{aligned} \lim_{n\to\infty} \frac{1}{2^n} \sum_{k=0}^{n} \binom{n}{k} \frac{1}{k+1} (k+1)^2 &= \lim_{n\to\infty} \left[\frac{1}{2^n} \sum_{k=0}^{n} \binom{n}{k} k + \frac{1}{2^n} \sum_{k=0}^{n} \binom{n}{k} \right] \\ &= \lim_{n\to\infty} \left(\frac{n2^{n-1}}{2^n} + 1 \right) = \infty. \end{aligned}$$

Hence, $xy \notin \breve{c}$. □

2.2.2 The Alpha-, Beta- and Gamma-duals of the Spaces $\breve{\ell}_\infty$, $\breve{c}$ and $\breve{c}_0$

In this subsection, we determine the alpha-, beta- and gamma-duals of the spaces $\breve{\ell}_\infty$, $\breve{c}$ and $\breve{c}_0$ of Euler-Cesàro bounded, convergent and null difference sequences.

Theorem 2.2.9. *The alpha-dual of the spaces $\breve{\ell}_\infty$, $\breve{c}$ and $\breve{c}_0$ is the set*

$$d_1 := \left\{ (a_n) \in \omega : \sup_{K \in \mathcal{F}} \sum_n \left| \sum_{k \in K} \sum_{j=k}^{n} \binom{j}{k} (-1)^{j-k} 2^k (2j-k+1) a_n \right| < \infty \right\}.$$

Proof. Let us define the matrix $M = (m_{nk})$ via $a = (a_n) \in \omega$ by

$$m_{nk} = \begin{cases} \sum_{j=k}^{n} \binom{j}{k} (-1)^{j-k} 2^k (2j-k+1) a_n & , \quad 0 \leq k \leq n, \\ 0 & , \quad k > n \end{cases}$$

for all $k, n \in \mathbb{N}$. Since the relation (2.3) holds, one can immediately derive that

$$a_n x_n = \sum_{k=0}^{n} \sum_{j=k}^{n} \binom{j}{k} (-1)^{j-k} 2^k (2j-k+1) a_n y_k = (My)_n \tag{2.6}$$

for all $n \in \mathbb{N}$. Therefore, we conclude by (2.6) that $a = (a_n) \in \lambda^\alpha$ if and only if $M \in (\mu : \ell_1)$. Then, we derive by Part (a) of Lemma 2.1.6 that

$$\sup_{K \in \mathcal{F}} \sum_n \left| \sum_{k \in K} \sum_{j=k}^{n} \binom{j}{k} (-1)^{j-k} 2^k (2j-k+1) a_n \right| < \infty$$

which leads to the desired result that $\lambda^\alpha = d_1$. □

Theorem 2.2.10. *Define the set d_2 by*

$$d_2 := \left\{ a = (a_i) \in \omega : \left(\sum_{k=0}^{n} \sum_{i=k}^{n} \sum_{j=k}^{i} \binom{j}{k} (-1)^{j-k} 2^k (2j-k+1) a_i \right) \in c \right\}.$$

Then, $\breve{c}^\beta = d_2$.

Proof. Define the triangle $B(a) = (b_{nk})$ by

$$b_{nk} = \begin{cases} a_n d_{nk} & , \quad 0 \leq k \leq n, \\ 0 & , \quad k > n \end{cases} \tag{2.7}$$

for all $k, n \in \mathbb{N}$, where $a = (a_n) \in \omega$ and d_{nk} is defined as in (2.4). Let $\widetilde{A} = (\widetilde{a}_{nk})$ be the composition of the matrices S and $B(a)$, i.e.,

$$\widetilde{a}_{nk} = \begin{cases} \sum_{i=k}^{n} \sum_{j=k}^{i} \binom{j}{k}(-1)^{j-k}2^k(2j-k+1)a_i & , \quad 0 \leq k \leq n \\ 0 & , \quad k > n \end{cases} \tag{2.8}$$

for all $k, n \in \mathbb{N}$. Therefore, we have by the relation (2.6) that

$$\sum_{k=0}^{n} a_k x_k = \sum_{k=0}^{n} \sum_{i=k}^{n} \sum_{j=k}^{i} \binom{j}{k}(-1)^{j-k}2^k(2j-k+1)a_i y_k \tag{2.9}$$

for all $n \in \mathbb{N}$. Then, (2.9) gives the fact that $ax \in cs$ whenever $x \in \breve{c}$ if and only if $\widetilde{A} \in (c : c)$. This means that $a \in \breve{c}^{\beta}$ if and only if $\widetilde{A} \in (c : c)$. Hence, we have by Part (c) of Lemma 2.1.6 that the following conditions are satisfied:

$$\sup_{n\in\mathbb{N}} \sum_{k=0}^{n} |\widetilde{a}_{nk}| = \sup_{n\in\mathbb{N}} \sum_{k=0}^{n} \left| \sum_{i=k}^{n} \sum_{j=k}^{i} \binom{j}{k}(-1)^{j-k}2^k(2j-k+1)a_i \right| < \infty,$$

$$\exists(\alpha_k) \in \omega \ni \lim_{n\to\infty} \widetilde{a}_{nk} = \lim_{n\to\infty} \sum_{i=k}^{n} \sum_{j=k}^{i} \binom{j}{k}(-1)^{j-k}2^k(2j-k+1)a_i = \alpha_k$$

$$\exists\alpha \in \mathbb{C} \ni \lim_{n\to\infty} \sum_{k=0}^{n} \widetilde{a}_{nk} = \lim_{n\to\infty} \sum_{k=0}^{n} \sum_{i=k}^{n} \sum_{j=k}^{i} \binom{j}{k}(-1)^{j-k}2^k(2j-k+1)a_i = \alpha$$

which give that

$$\left\{ \sum_{k=0}^{n} \sum_{i=k}^{n} \sum_{j=k}^{i} \binom{j}{k}(-1)^{j-k}2^k(2j-k+1)a_i \right\}_{n\in\mathbb{N}} \in c.$$

That is to say that the beta-dual of the space $\breve{c}$ is the set d_2. □

Theorem 2.2.11. *$\breve{c}_0^{\gamma} = \breve{c}^{\gamma} = (\breve{\ell}_{\infty})^{\gamma} = d_2$, where the set d_2 is defined by (2.25).*

Proof. This is similar to the proof of Theorem 2.2.10 with Part (b) instead of Part (c) of Lemma 2.10. So, we omit details. □

2.2.3 Matrix Transformations Related to the Sequence Space $\breve{c}$

We characterize some matrix classes from the space $\breve{c}$ into the classical sequence spaces ℓ_{∞}, c and c_0.

Theorem 2.2.12. *$A = (a_{nk}) \in (\breve{c} : \ell_{\infty})$ if and only if*

$$A_n \in \breve{c}^{\beta} \quad \textit{for each} \quad n \in \mathbb{N}, \tag{2.10}$$

$$\sup_{n\in\mathbb{N}}\sum_k\left|\sum_{i=k}^{\infty}\sum_{j=k}^{i}\binom{j}{k}(-1)^{j-k}2^k(2j-k+1)a_{ni}\right|<\infty. \tag{2.11}$$

Proof. Suppose that $A=(a_{nk})\in(\breve{c}:\ell_\infty)$ and $x=(x_k)\in\breve{c}$. Since Ax exists and belongs to the space ℓ_∞, the necessity of the condition (2.10) is obvious.

Consider the following equality derived from the mth partial sum of the series $\sum_k a_{nk}x_k$:

$$\sum_{k=0}^{m}a_{nk}x_k=\sum_{k=0}^{m}\sum_{i=k}^{m}\sum_{j=k}^{i}\binom{j}{k}(-1)^{j-k}2^k(2j-k+1)a_{ni}y_k \tag{2.12}$$

for all $m,n\in\mathbb{N}$. Therefore, by letting $m\to\infty$ in the equality (2.12) it is immediate that

$$\sum_k a_{nk}x_k=\sum_k\sum_{i=k}^{\infty}\sum_{j=k}^{i}\binom{j}{k}(-1)^{j-k}2^k(2j-k+1)a_{ni}y_k \tag{2.13}$$

for all $n\in\mathbb{N}$. Now, define the matrix $F=(f_{nk})$ by

$$f_{nk}=\sum_{i=k}^{\infty}\sum_{j=k}^{i}\binom{j}{k}(-1)^{j-k}2^k(2j-k+1)a_{ni} \tag{2.14}$$

for all $k,n\in\mathbb{N}$. Then, $Ax=Fy$ which gives that $F\in(c:\ell_\infty)$. Then, F satisfies the condition (2.4) which is equivalent to (2.11).

Conversely, suppose that the conditions (2.10) and (2.11) hold, and take $x=(x_k)\in\breve{c}$. Then, (2.10) implies the existence of Ax and since the spaces $\breve{c}$ and c are isomorphic we have $y\in c$. Therefore, (2.13) gives with (2.11) that

$$\begin{aligned}\|Ax\|_\infty &= \sup_{n\in\mathbb{N}}\left|\sum_k a_{nk}x_k\right|\\ &\le \sup_{n\in\mathbb{N}}\sum_k\left|\sum_{i=k}^{n}\sum_{j=k}^{i}\binom{j}{k}(-1)^{j-k}2^k(2j-k+1)a_{ni}y_k\right|\\ &\le \|y\|_\infty\left[\sup_{n\in\mathbb{N}}\sum_k\left|\sum_{i=k}^{n}\sum_{j=k}^{i}\binom{j}{k}(-1)^{j-k}2^k(2j-k+1)a_{ni}\right|\right]\\ &< \infty.\end{aligned}$$

Hence, $A\in(\breve{c}:\ell_\infty)$.

This completes the proof. □

Theorem 2.2.13. *$A=(a_{nk})\in(\breve{c}:c)$ if and only if the conditions (2.10) and (2.11) hold, and*

$$\exists\beta_k\in\mathbb{C}\ \text{ such that }\ \lim_{n\to\infty}f_{nk}=\beta_k \text{ for all } k\in\mathbb{N}, \tag{2.15}$$

$$\exists \beta \in \mathbb{C} \quad \text{such that} \quad \lim_{n\to\infty} \sum_k f_{nk} = \beta, \tag{2.16}$$

where f_{nk} is defined by (2.14) as in the proof of Theorem 2.2.12.

Proof. Suppose that the conditions (2.10), (2.11), (2.15) and (2.16) hold, and take any $x = (x_k) \in \check{c}$. The condition (2.10) implies the existence of A-transform of x. Then, one can derive by using the hypothesis (2.11) that

$$\sum_{k=0}^{m} |\beta_k| \leq \sup_{n\in\mathbb{N}} \sum_k |f_{nk}| < \infty$$

for all $m \in \mathbb{N}$. Hence, $(\beta_k) \in \ell_1$ which implies that $(\beta_k y_k) \in \ell_1$. Therefore, by taking in mind the relation (2.13) we derive for all $n \in \mathbb{N}$ that

$$\sum_k a_{nk} x_k = \sum_k (f_{nk} - \beta_k) y_k + \sum_k \beta_k y_k.$$

Then, we have by letting $n \to \infty$ that

$$\lim_{n\to\infty} (Ax)_n = \sum_k \beta_k y_k. \tag{2.17}$$

Since $(\beta_k y_k) \in \ell_1$, (2.17) gives that $Ax \in c$, that is, $A \in (\check{c} : c)$.

Conversely, suppose that $A = (a_{nk}) \in (\check{c} : c)$ and take $x \in \check{c}$. Since the inclusion relation $c \subset \ell_\infty$ holds, the necessity of the conditions (2.10) and (2.11) follows from Theorem 2.2.12.

Now, consider the convergent sequences $u = (u_k) = \left\{u_k^{(n)}\right\}_{k\in\mathbb{N}}$ defined by (2.5) and $x = (x_k) = \left\{\sum_{j=0}^{k} (-1)^{k-j} D_{k-j} T_j\right\}$. Since A-transforms of u and x exist, and belong to the space c by the hypothesis, one can see that $Au = \left\{\sum_{j=k}^{\infty} (-1)^{j-k} D_{j-k} T_k a_{nj}\right\}_{n\in\mathbb{N}} \in c$ and $Ax = \left(\sum_k f_{nk}\right)_{n\in\mathbb{N}} \in c$ which shows the necessity of the conditions (2.15) and (2.16), respectively.

This completes the proof. □

Corollary 2.2.14. *$A = (a_{nk}) \in (\check{c} : c_0)$ if and only if (2.10), (2.11) hold, and (2.15) also holds with $\beta_k = 0$ for all $k \in \mathbb{N}$.*

Now, we can give the theorem characterizing the class of matrix transformations from a given sequence space ν to the Nörlund space λ.

Theorem 2.2.15. *Suppose that ν be any given sequence space and the infinite matrices $A = (a_{nk})$ and $\widetilde{D} = (\widetilde{d}_{nk})$ are connected with the relation*

$$\widetilde{d}_{nk} = \sum_{j=0}^{n} \sum_{i=0}^{n} \frac{\binom{n}{i} a_{ik}}{2^n (i+1)} \quad \text{for all} \quad k, n \in \mathbb{N}.$$

Then, $A \in (\nu : \lambda)$ if and only if $\widetilde{D} \in (\nu : \mu)$.

Proof. Let $x = (x_k) \in \nu$. Consider the following equality

$$\begin{aligned}\sum_{k=0}^{m} \widetilde{d}_{nk}x_k &= \sum_{k=0}^{m}\sum_{j=0}^{n}\frac{1}{2^n}\sum_{i=0}^{n}\binom{n}{i}\frac{1}{i+1}a_{ik}x_k \\ &= \sum_{j=0}^{n}\binom{n}{j}\frac{1}{2^n}\sum_{i=0}^{n}\binom{n}{i}\frac{1}{i+1}\sum_{k=0}^{m}a_{ik}x_k \quad \text{for all } m,n \in \mathbb{N}.\end{aligned} \tag{2.18}$$

Then, by letting $m \to \infty$ in (2.18) it is immediate that $(\widetilde{D}x)_n = \{\widetilde{B}(Ax)\}_n$ for all $n \in \mathbb{N}$. Therefore, one can easily see that $\widetilde{D}x \in \mu$ whenever $x \in \nu$ if and only if $Ax \in \lambda$ whenever $x \in \nu$.

This completes the proof. □

By combining Theorems 2.2.12, 2.2.13, 2.2.15 and Corollary 2.2.14, the following results are derived on the characterization of some matrix classes:

Corollary 2.2.16. *Let $L = (l_{nk})$ be an infinite matrix over the complex field. Then, the following statements hold:*

(i) $L \in (\check{c} : bs)$ *if and only if (2.10) and (2.11) hold with r_{nk} instead of a_{nk}, where $r_{nk} = \sum_{j=0}^{n} l_{jk}$ for all $k,n \in \mathbb{N}$.*

(ii) $L \in (\check{c} : bv_\infty)$ *if and only if (2.10) and (2.11) hold with r_{nk} instead of a_{nk}, where $r_{nk} = l_{nk} - l_{n-1,k}$ for all $k,n \in \mathbb{N}$.*

(iii) $L \in (\check{c} : X_\infty)$ *if and only if (2.10) and (2.11) hold with r_{nk} instead of a_{nk}, where $r_{nk} = \sum_{j=0}^{n} l_{jk}/(n+1)$ for all $k,n \in \mathbb{N}$.*

(iv) $L \in (\check{c} : r_\infty^q)$ *if and only if (2.10) and (2.11) hold with r_{nk} instead of a_{nk}, where $r_{nk} = \sum_{j=0}^{n} q_j l_{jk}/Q_n$ for all $k,n \in \mathbb{N}$.*

(v) $L \in (\check{c} : a_\infty^r)$ *if and only if (2.10) and (2.11) hold with r_{nk} instead of a_{nk}, where $r_{nk} = \sum_{j=0}^{n}(1+r^j)l_{jk}/(1+n)$ for all $k,n \in \mathbb{N}$.*

(vi) $L \in (\check{c} : e_\infty^r)$ *if and only if (2.10) and (2.11) hold with r_{nk} instead of a_{nk}, where $r_{nk} = \sum_{j=0}^{n}\binom{n}{j}(1-r)^{n-j}r^j l_{jk}$ for all $k,n \in \mathbb{N}$.*

Corollary 2.2.17. *Let $L = (l_{nk})$ be an infinite matrix over the complex field. Then, the following statements hold:*

(i) $L \in (\check{c} : cs)$ *if and only if (2.10), (2.11), (2.15) and (2.16) hold with r_{nk} instead of a_{nk}, where $r_{nk} = \sum_{j=0}^{n} l_{jk}$ for all $k,n \in \mathbb{N}$.*

(ii) $L \in (\check{c} : c(\Delta))$ *if and only if (2.10), (2.11), (2.15) and (2.16) hold with r_{nk} instead of a_{nk}, where $r_{nk} = l_{nk} - l_{n-1,k}$ for all $k,n \in \mathbb{N}$.*

(iii) $L \in (\check{c} : \widetilde{c})$ *if and only if (2.10), (2.11), (2.15) and (2.16) hold with r_{nk} instead of a_{nk}, where $r_{nk} = \sum_{j=0}^{n} l_{jk}/(n+1)$ for all $k,n \in \mathbb{N}$.*

(iv) $L \in (\breve{c} : r_c^q)$ *if and only if (2.10), (2.11), (2.15) and (2.16) hold with* r_{nk} *instead of* a_{nk}*, where* $r_{nk} = \sum_{j=0}^{n} q_j l_{jk}/Q_n$ *for all* $k, n \in \mathbb{N}$.

(v) $L \in (\breve{c} : a_c^r)$ *if and only if (2.10), (2.11), (2.15) and (2.16) hold with* r_{nk} *instead of* a_{nk}*, where* $r_{nk} = \sum_{j=0}^{n}(1 + r^j)l_{jk}/(1 + n)$ *for all* $k, n \in \mathbb{N}$.

(vi) $L \in (\breve{c} : e_c^r)$ *if and only if (2.10), (2.11), (2.15) and (2.16) hold with* r_{nk} *instead of* a_{nk}*, where* $r_{nk} = \sum_{j=0}^{n} \binom{n}{j}(1 - r)^{n-j}r^j l_{jk}$ *for all* $k, n \in \mathbb{N}$.

Corollary 2.2.18. *Let* $L = (l_{nk})$ *be an infinite matrix over the complex field. Then, the following statements hold:*

(i) $L \in (\breve{c} : cs_0)$ *if and only if the conditions (2.10), (2.11) and (2.15) of Corollary 2.2.14 hold with* r_{nk} *instead of* a_{nk}*, where* cs_0 *denotes the space of all series converging to zero and* $r_{nk} = \sum_{j=0}^{n} l_{jk}$ *for all* $k, n \in \mathbb{N}$.

(ii) $L \in (\breve{c} : c_0(\nabla))$ *if and only if the conditions (2.10), (2.11) and (2.15) of Corollary 2.2.14 hold with* r_{nk} *instead of* a_{nk}*, where* $r_{nk} = l_{nk} - l_{n-1,k}$ *for all* $k, n \in \mathbb{N}$.

(iii) $L \in (\breve{c} : \widetilde{c}_0)$ *if and only if the conditions (2.10), (2.11) and (2.15) of Corollary 2.2.14 hold with* r_{nk} *instead of* a_{nk}*, where* $r_{nk} = \sum_{j=0}^{n} l_{jk}/(n+1)$ *for all* $k, n \in \mathbb{N}$.

(iv) $L \in (\breve{c} : r_0^q)$ *if and only if the conditions (2.10), (2.11) and (2.15) of Corollary 2.2.14 hold with* r_{nk} *instead of* a_{nk}*, where* $r_{nk} = \sum_{j=0}^{n} q_j l_{jk}/Q_n$ *for all* $k, n \in \mathbb{N}$.

(v) $L \in (\breve{c} : a_0^r)$ *if and only if the conditions (2.10), (2.11) and (2.15) of Corollary 2.2.14 hold with* r_{nk} *instead of* a_{nk}*, where* $r_{nk} = \sum_{j=0}^{n}(1 + r^j)l_{jk}/(1 + n)$ *for all* $k, n \in \mathbb{N}$.

(vi) $L \in (\breve{c} : e_0^r)$ *if and only if the conditions (2.10), (2.11) and (2.15) of Corollary 2.2.14 hold with* r_{nk} *instead of* a_{nk}*, where* $r_{nk} = \sum_{j=0}^{n} \binom{n}{j}(1 - r)^{n-j}r^j l_{jk}$ *for all* $k, n \in \mathbb{N}$.

2.2.4 The Euler-Cesàro Difference Spaces of Absolutely p-Summable Sequences

In this subsection, following Başar [22, 23], we investigate the space $\widetilde{\ell}_p$ of Euler-Cesàro absolutely p-summable difference sequences. Also, we present an inclusion theorem together with the Schauder basis of the space $\widetilde{\ell}_p$. We characterize the classes $(\widetilde{\ell}_p : \ell_\infty)$, $(\widetilde{\ell}_p : c)$ and $(\widetilde{\ell}_p : c_0)$ of infinite matrices, and give the characterization of some other classes of matrix transformations from the space $\widetilde{\ell}_p$ to the Euler, Riesz, difference, etc., sequence spaces, by means of a basic lemma.

Now, we introduce the space $\widetilde{\ell_p}$ of Euler-Cesàro absolutely p-summable difference sequences as the set of all sequences whose $\widetilde{B}$-transforms are in the space ℓ_p with $0 < p < \infty$, i.e.,

$$\widetilde{\ell_p} := \left\{ x = (x_k) \in \omega : \sum_n \left| \sum_{k=0}^{n} \frac{\binom{n}{k}}{2^n(k+1)} x_k \right|^p < \infty \right\}, \quad (0 < p < \infty).$$

Of course, the space $\widetilde{\ell_p}$ can be seen as the domain of the composition of E_1 and C in the difference space bv_p introduced in the cases $1 \leq p \leq \infty$ and $0 < p < 1$ by Başar and Altay [25] and Altay and Başar [10], respectively. With the notation of (1.3), the space $\widetilde{\ell_p}$ can be redefined by $\widetilde{\ell_p} = (\ell_p)_{\widetilde{B}}$.

Theorem 2.2.19. *The set $\widetilde{\ell_p}$ is a linear space with the coordinatewise addition and scalar multiplication of sequences. In the case $0 < p \leq 1$, $\widetilde{\ell_p}$ is a complete p-normed space with the p-norm $\|x\|_{\widetilde{p}} = \|\widetilde{B}x\|_p$ and in the case $1 < p < \infty$, $\widetilde{\ell_p}$ is a BK-space with the norm $\|x\|_{\widetilde{p}} = \|\widetilde{B}x\|_p$.*

Proof. Let $0 < p < \infty$. It is a routine verification that the set $\widetilde{\ell_p}$ is a linear space with the coordinatewise addition and scalar multiplication of sequences. So, we omit details.

Additionally, in the cases $0 < p \leq 1$ and $1 < p < \infty$ since ℓ_p is a p-normed space with respect to its usual p-norm $\|\cdot\|_{\widetilde{p}}$ and is a BK-space with respect to its usual norm $\|\cdot\|_p$, and $\widetilde{B}$ is a triangle matrix, Theorem 4.3.2 of Wilansky [135, p. 61] gives the fact that $\widetilde{\ell_p}$ is a complete p-normed space and is a BK-space, respectively.

This completes the proof. □

With the notation of (2.2), since the transformation T defined from $\widetilde{\ell_p}$ to ℓ_p by $x \mapsto y = Tx = \widetilde{B}x$ is a linear bijection, we have the following:

Corollary 2.2.20. *In the cases $0 < p \leq 1$ and $1 < p < \infty$, the sequence space $\widetilde{\ell_p}$ is linearly p-norm isomorphic and norm isomorphic to the space ℓ_p, respectively.*

Now, we give the following inclusion relation concerning the space $\widetilde{\ell_p}$.

Theorem 2.2.21. *Let $0 < p < \infty$. Then, the inclusion $\ell_p \subset \widetilde{\ell_p}$ strictly holds.*

Proof. To prove the validity of the inclusion $\ell_p \subset \widetilde{\ell_p}$, it suffices to show the existence of a number $K > 0$ such that $\|x\|_{\widetilde{p}} \leq K\|x\|_p$ for every $x \in \ell_p$.

Let us take any $x = (x_k) \in \ell_p$ with $0 < p \leq 1$. Then, one can see by taking into account the fact $|x_k| \leq \|x\|_p$ for all $k \in \mathbb{N}$ that

$$\|x\|_{\widetilde{p}} = \|\widetilde{B}x\|_p = \sum_n \left| \sum_{k=0}^{n} \frac{\binom{n}{k}}{2^n(k+1)} x_k \right|^p \tag{2.19}$$

$$
\begin{aligned}
&\leq \sum_n \left[\sum_{k=0}^{n} \frac{\binom{n}{k}}{2^n(k+1)} |x_k| \right]^p \\
&\leq \|x\|_p \sum_n \left[\frac{2^{n+1}-1}{2^n(n+1)} \right]^p \\
&< \|x\|_p \sum_n \left(\frac{2}{n+1} \right)^p < \infty.
\end{aligned}
$$

Thus, (2.19) shows that $\|\widetilde{x}\|_p < \|x\|_p \|\{2/(n+1)\}\|_p < \infty$. That is to say that the inclusion $\ell_p \subset \widetilde{\ell_p}$ holds.

Let us consider the sequence $u = (u_k)$ defined by $u_k = (-1)^k$ for all $k \in \mathbb{N}$. Then, since $\|\widetilde{u}\|_p = \|\widetilde{B}u\|_p < \|\{2/(n+1)\}\|_p < \infty$, it is immediate that u is in $\widetilde{\ell_p} \setminus \ell_p$. Because there is at least one sequence in $\widetilde{\ell_p} \setminus \ell_p$, the inclusion $\ell_p \subset \widetilde{\ell_p}$ is strict.

Let us take any $x = (x_k) \in \ell_p$ with $1 < p < \infty$. Then, one can see by taking into account $|x_k| \leq \|x\|_p$ for all $k \in \mathbb{N}$ that

$$
\begin{aligned}
\left(\|\widetilde{x}\|_p\right)^p = \left(\|\widetilde{B}x\|_p\right)^p &= \sum_n \left| \sum_{k=0}^{n} \frac{\binom{n}{k}}{2^n(k+1)} x_k \right|^p \qquad (2.20) \\
&\leq \sum_n \left[\sum_{k=0}^{n} \frac{\binom{n}{k}}{2^n(k+1)} |x_k| \right]^p \\
&\leq (\|x\|_p)^p \sum_n \left[\frac{2^{n+1}-1}{2^n(n+1)} \right]^p \\
&< (\|x\|_p)^p \sum_n \left(\frac{2}{n+1} \right)^p < \infty.
\end{aligned}
$$

Thus, (2.20) shows that $\|\widetilde{x}\|_p < \|x\|_p \, \|\{2/(n+1)\}\|_p < \infty$. That is to say that the inclusion $\ell_p \subset \widetilde{\ell_p}$ holds.

Let us consider the sequence $u = (u_k)$ defined by $u_k = (-1)^k$ for all $k \in \mathbb{N}$. Then, it is immediate that since $\|\widetilde{u}\|_p = \|\widetilde{B}u\|_p < \|\{2/(n+1)\}\|_p < \infty$ we immediately observe that u is in $\widetilde{\ell_p}$ but is not in ℓ_p. Because there is at least one sequence in $\widetilde{\ell_p} \setminus \ell_p$, the inclusion $\ell_p \subset \widetilde{\ell_p}$ is strict.

This completes the proof. □

As a direct consequence of Remark 2.1.4, we have:

Corollary 2.2.22. *Let $0 < p < \infty$ and $\alpha_n = (\widetilde{B}x)_n$ for all $n \in \mathbb{N}$. Define the sequence $u^{(n)} = \left\{u_k^{(n)}\right\}_{k\in\mathbb{N}}$ in the space $\widetilde{\ell}_p$ by*

$$u_k^{(n)} = \begin{cases} \sum\limits_{j=k}^{n} \binom{j}{k}(-1)^{j-k}2^k(2j-k+1) & , \quad 0 \le k \le n, \\ \\ 0 & , \quad k > n \end{cases} \tag{2.21}$$

for every fixed $n \in \mathbb{N}$. The sequence $\{u^{(n)}\}_{n\in\mathbb{N}}$ is a basis for the space $\widetilde{\ell}_p$ and any $x \in \widetilde{\ell}_p$ has a unique representation of the form $x = \sum_n \alpha_n u^{(n)}$.

2.2.5 The Alpha-, Beta- and Gamma-duals of the Space $\widetilde{\ell}_p$

In this subsection, we determine the alpha-, beta- and gamma-duals of the space $\widetilde{\ell}_p$ of Euler-Cesàro absolutely p-summable difference sequences.

Now, we start with the following lemma due to Steieglitz and Tietz [127] which is needed in proving the next theorems.

Lemma 2.2.23. *Let $A = (a_{nk})$ be an infinite matrix over the complex field. Then, the following statements hold:*

(a) $A \in (\ell_p : \ell_1)$ if and only if

$$\text{In the case } 0 < p \le 1, \quad \sup_{k\in\mathbb{N}} \sum_n |a_{nk}| < \infty,$$

$$\text{In the case } 1 < p < \infty, \quad \sup_{K\in\mathcal{F}} \sum_k \left|\sum_{n\in K} a_{nk}\right|^q < \infty.$$

(b) $A \in (\ell_p : c)$ if and only if (2.5) holds and

$$\text{In the case } 0 < p \le 1, \quad \sup_{k,n\in\mathbb{N}} |a_{nk}| < \infty, \tag{2.22}$$

$$\text{In the case } 1 < p < \infty, \quad \sup_{n\in\mathbb{N}} \sum_k |a_{nk}|^q < \infty. \tag{2.23}$$

(c) $A \in (\ell_p : \ell_\infty)$ if and only if in the case $0 < p \le 1$, (2.22) holds and in the case $1 < p < \infty$, (2.23) holds.

Theorem 2.2.24. *Define the sets D_1 and D_2 by*

$$D_1 := \left\{(a_n) \in \omega : \sup_{k\in\mathbb{N}} \sum_n \left|\sum_{j=k}^{n} \binom{j}{k}(-1)^{j-k}2^k(2j-k+1)a_n\right| < \infty\right\},$$

$$D_2 := \left\{(a_i) \in \omega : \sup_{K\in\mathcal{F}} \sum_{i=0}^{n} \left|\sum_{k\in K}\sum_{j=k}^{n} \binom{j}{k}(-1)^{j-k}2^k(2j-k+1)a_i\right|^q < \infty\right\}.$$

Then, $[\widetilde{\ell_p}]^\alpha = \begin{cases} D_1 & , \text{ if } 0 < p \leq 1, \\ D_2 & , \text{ if } 1 < p < \infty. \end{cases}$

Proof. Let us define the matrix $M = (m_{nk})$ via $a = (a_n) \in \omega$ by

$$m_{nk} = \begin{cases} \sum\limits_{j=k}^{n} \binom{j}{k}(-1)^{j-k}2^k(2j-k+1)a_n & , \quad 0 \leq k \leq n, \\ 0 & , \quad k > n \end{cases}$$

for all $k, n \in \mathbb{N}$. One can easily obtain by using the relation (2.2) that

$$a_n x_n = \sum_{k=0}^{n}\sum_{j=k}^{n} \binom{j}{k}(-1)^{j-k}2^k(2j-k+1)a_n y_k = (My)_n \tag{2.24}$$

for all $n \in \mathbb{N}$. Therefore, we conclude from (2.24) that $a = (a_n) \in [\widetilde{\ell_p}]^\alpha$ if and only if $M \in (\ell_p : \ell_1)$. Then, we derive by Part (a) of Lemma 2.2.23 for the cases $0 < p \leq 1$ and $1 < p < \infty$ that

$$\sup_{k \in \mathbb{N}} \sum_n \left| \sum_{j=k}^{n} \binom{j}{k}(-1)^{j-k}2^k(2j-k+1)a_n \right| < \infty,$$

$$\sup_{K \in \mathcal{F}} \sum_{i=0}^{n} \left| \sum_{k \in K}\sum_{j=k}^{n} \binom{j}{k}(-1)^{j-k}2^k(2j-k+1)a_i \right|^q < \infty$$

which lead us to the desired result. □

Theorem 2.2.25. *Define the sets D_3, D_4 and D_5 by*

$$D_3 := \left\{(a_i) \in \omega : \sup_{k,n \in \mathbb{N}} \left| \sum_{i=k}^{n}\sum_{j=k}^{i} \binom{j}{k}(-1)^{j-k}2^k(2j-k+1)a_i \right| < \infty \right\},$$

$$D_4 := \left\{(a_i) \in \omega : \exists \alpha_k \in \mathbb{C} \ni \lim_{n\to\infty} \sum_{i=k}^{n}\sum_{j=k}^{i} \binom{j}{k}(-1)^{j-k}2^k(2j-k+1)a_i = \alpha_k \right\},$$

$$D_5 := \left\{(a_i) \in \omega : \sup_{n \in \mathbb{N}} \sum_{k=0}^{n} \left| \sum_{i=k}^{n}\sum_{j=k}^{i} \binom{j}{k}(-1)^{j-k}2^k(2j-k+1)a_i \right|^q < \infty \right\}.$$

Then, $[\widetilde{\ell_p}]^\beta = \begin{cases} D_3 \cap D_4 & , \text{ if } 0 < p \leq 1, \\ D_5 & , \text{ if } 1 < p < \infty. \end{cases}$

Proof. Consider the triangles $B(a) = (b_{nk})$ and $\widetilde{A} = (\widetilde{a}_{nk})$ defined by (2.7) and (2.8), respectively. Therefore, we again have the relation (2.9) which leads to the fact that $ax \in cs$ whenever $x \in \widetilde{\ell_p}$ if and only if $\widetilde{A} \in (\ell_p : c)$. Hence, we have by Part (b) of Lemma 2.2.23 that the following conditions are satisfied:

$$\sup_{k,n \in \mathbb{N}} |\widetilde{a}_{nk}| = \sup_{k,n \in \mathbb{N}} \left| \sum_{i=k}^{n}\sum_{j=k}^{i} \binom{j}{k}(-1)^{j-k}2^k(2j-k+1)a_i \right| < \infty,$$

$$\exists(\alpha_k) \in \omega \ni \lim_{n\to\infty} \widetilde{a}_{nk} = \lim_{n\to\infty} \sum_{i=k}^{n}\sum_{j=k}^{i} \binom{j}{k}(-1)^{j-k}2^k(2j-k+1)a_i = \alpha_k$$

and

$$\sup_{n\in\mathbb{N}} \sum_{k=0}^{n} |\widetilde{a}_{nk}|^q = \sup_{n\in\mathbb{N}} \sum_{k=0}^{n} \left| \sum_{i=k}^{n}\sum_{j=k}^{i} \binom{j}{k}(-1)^{j-k}2^k(2j-k+1)a_i \right|^q < \infty,$$

$$\exists\alpha \in \mathbb{C} \ni \lim_{n\to\infty} \sum_{k=0}^{n} \widetilde{a}_{nk} = \lim_{n\to\infty} \sum_{k=0}^{n}\sum_{i=k}^{n}\sum_{j=k}^{i} \binom{j}{k}(-1)^{j-k}2^k(2j-k+1)a_i = \alpha$$

which give that the beta-dual of the space $\widetilde{\ell}_p$ is the set $D_3 \cap D_4$ if $0 < p \leq 1$ and is the set D_5 if $1 < p < \infty$, as asserted. □

Theorem 2.2.26. *The gamma-dual of the space $\widetilde{\ell}_p$ is the set D_3 if $0 < p \leq 1$ and is the set D_5 if $1 < p < \infty$.*

Proof. This is similar to the proof of Theorem 2.2.25 with Part (c) instead of Part (b) of Lemma 2.2.23. To avoid the repetition of the similar statements, we omit details. □

2.2.6 Matrix Transformations on the Sequence Space $\widetilde{\ell}_p$

In this subsection, we desire to characterize some classes of matrix transformations on the space $\widetilde{\ell}_p$. We directly prove the theorems characterizing the classes $(\widetilde{\ell}_p : \ell_\infty)$ and $(\widetilde{\ell}_p : c)$, and derive the characterizations of some other classes from them by means of Lemma 2.2.30.

Theorem 2.2.27. $A = (a_{nk}) \in (\widetilde{\ell}_p : \ell_\infty)$ *if and only if*

(a) In the case $0 < p \leq 1$,

$$\sup_{k,n\in\mathbb{N}} \left| \sum_{i=k}^{\infty}\sum_{j=k}^{i} \binom{j}{k}(-1)^{j-k}2^k(2j-k+1)a_{ni} \right| < \infty. \tag{2.25}$$

(b) In the case $1 < p < \infty$,

$$\sup_{n\in\mathbb{N}} \sum_{k} \left| \sum_{i=k}^{\infty}\sum_{j=k}^{i} \binom{j}{k}(-1)^{j-k}2^k(2j-k+1)a_{ni} \right|^q < \infty. \tag{2.26}$$

Proof. (a) Suppose that $A = (a_{nk}) \in (\widetilde{\ell}_p : \ell_\infty)$ and $x = (x_k) \in \widetilde{\ell}_p$ with $0 < p \leq 1$. Then, Ax exists and belongs to the space ℓ_∞. Therefore, one can conclude for the sequence $u^{(n)} = \left\{u_k^{(n)}\right\}_{k\in\mathbb{N}}$, defined by (2.21), in the space

$\widetilde{\ell_p}$ that

$$Au^{(n)} = \left(\sum_{i=k}^{\infty} \sum_{j=k}^{i} \binom{j}{k} (-1)^{j-k} 2^k (2j-k+1) a_{nj} \right) \in \ell_\infty$$

for each $n \in \mathbb{N}$. Hence, the condition (2.25) is necessary.

Conversely, suppose that (2.25) holds and take any $x = (x_k) \in \widetilde{\ell_p}$. Then, $A_n \in [\widetilde{\ell_p}]^\beta$ for each $n \in \mathbb{N}$ which leads to the existence of Ax. Let $n \in \mathbb{N}$ be fixed. By following the similar way used in the first part of the proof of Theorem 2.2.12, we again have the relation (2.13) which means that

$$(Ax)_n = \sum_k a_{nk} x_k = \sum_k f_{nk} y_k = (Fy)_n \tag{2.27}$$

for all $n \in \mathbb{N}$, where the matrix $F = (f_{nk})$ is defined by (2.14). Therefore, since

$$\begin{aligned} \left| \sum_k a_{nk} x_k \right|^p &= \left| \sum_k \sum_{i=k}^{m} \sum_{j=k}^{i} \binom{j}{k} (-1)^{j-k} 2^k (2j-k+1) a_{ni} y_k \right|^p \\ &\le \left(\sum_k \left| \sum_{i=k}^{m} \sum_{j=k}^{i} \binom{j}{k} (-1)^{j-k} 2^k (2j-k+1) a_{ni} \right| |y_k| \right)^p \\ &\le \left(\sup_{k \in \mathbb{N}} \left| \sum_{i=k}^{m} \sum_{j=k}^{i} \binom{j}{k} (-1)^{j-k} 2^k (2j-k+1) a_{ni} \right| \right)^p \left(\sum_k |y_k| \right)^p \\ &\le \left(\sup_{k \in \mathbb{N}} \left| \sum_{i=k}^{m} \sum_{j=k}^{i} \binom{j}{k} (-1)^{j-k} 2^k (2j-k+1) a_{ni} \right| \right)^p \sum_k |y_k|^p \end{aligned}$$

we obtain by taking supremum over $n \in \mathbb{N}$ that

$$\begin{aligned} \|Ax\|_\infty &= \sup_{n \in \mathbb{N}} \left| \sum_k a_{nk} x_k \right| \\ &\le (\|y\|_p)^{1/p} \sup_{k,n \in \mathbb{N}} \left| \sum_{i=k}^{m} \sum_{j=k}^{i} \binom{j}{k} (-1)^{j-k} 2^k (2j-k+1) a_{ni} \right| \\ &< \infty. \end{aligned}$$

That is, $Ax \in \ell_\infty$, as desired.

(b) Suppose that $A = (a_{nk}) \in (\widetilde{\ell_p} : \ell_\infty)$ and $x = (x_k) \in \widetilde{\ell_p}$ with $1 < p < \infty$. Since Ax exists and belongs to the space ℓ_∞ by the hypothesis, $A_n \in [\widetilde{\ell_p}]^\beta$ for all $n \in \mathbb{N}$. This shows the necessity of the condition (2.26).

Conversely, suppose that the condition (2.26) holds and take $x = (x_k) \in \widetilde{\ell_p}$. Then, (2.26) implies the existence of Ax and since the spaces $\widetilde{\ell_p}$ and ℓ_p are norm isomorphic, we have $y = (y_k) \in \ell_p$. Consider the equality $(Ax)_n = (Fy)_n$ for all $n \in \mathbb{N}$ in (2.27) which gives that $Fy \in \ell_\infty$, where $F = (f_{nk})$ is defined by the relation (2.14). Therefore, one can see by applying Hölder's inequality to (2.27) with (2.26) that

$$\begin{aligned}
\|Ax\|_\infty &= \sup_{n\in\mathbb{N}}\left|\sum_k a_{nk}x_k\right| \\
&\leq \sup_{n\in\mathbb{N}}\sum_k\left|\sum_{i=k}^{\infty}\sum_{j=k}^{i}\binom{j}{k}(-1)^{j-k}2^k(2j-k+1)a_{ni}y_k\right| \\
&\leq (\|y\|_p)^p\left[\sup_{n\in\mathbb{N}}\sum_k\left|\sum_{i=k}^{\infty}\sum_{j=k}^{i}\binom{j}{k}(-1)^{j-k}2^k(2j-k+1)a_{ni}\right|^q\right] \\
&< \infty.
\end{aligned}$$

Hence, $A \in (\widetilde{\ell_p} : \ell_\infty)$.

This completes the proof. □

Theorem 2.2.28. *Let $0 < p < \infty$. Then, $A = (a_{nk}) \in (\widetilde{\ell_p} : c)$ if and only if*

(a) In the case $0 < p \leq 1$, the condition (2.25) holds and

$$\exists a_k \in \mathbb{C} \ \text{ such that } \ \lim_{n\to\infty} f_{nk} = a_k \text{ for all } k \in \mathbb{N}, \tag{2.28}$$

where f_{nk} is defined by (2.14) as in the proof of Theorem 2.2.27.

(b) In the case $1 < p < \infty$, (2.26) and (2.28) hold.

Proof. (a) Suppose that $A = (a_{nk}) \in (\widetilde{\ell_p} : c)$ with $0 < p \leq 1$. Then, Ax exists and is in the space c for all $x = (x_k) \in \widetilde{\ell_p}$. Since the inclusion $c \subset \ell_\infty$ holds, the necessity of the condition (2.25) follows from Theorem 2.2.27. The necessity of the condition (2.28) is immediate by taking the sequence $u = (u_k) = \left\{u_k^{(n)}\right\}_{k\in\mathbb{N}}$, defined by (2.21).

Conversely, suppose that the conditions (2.25) and (2.28) hold, and take any $x = (x_k) \in \widetilde{\ell_p}$. Then, since $A_n \in [\widetilde{\ell_p}]^\beta$ for each $n \in \mathbb{N}$, Ax exists. Then, by taking into account the relation (2.27) one can see that the conditions (2.25) and (2.28) correspond to (2.26) and (2.5) with f_{nk} instead of a_{nk}, respectively, where f_{nk} is given by (2.14). Hence, $Fy \in c$ which gives by (2.27) that $A \in (\widetilde{\ell_p} : c)$.

(b) Suppose that the conditions (2.26) and (2.28) hold, and take any $x = (x_k) \in \widetilde{\ell_p}$ with $1 < p < \infty$. The condition (2.26) implies the existence of A-transform of x. Therefore, one can derive by using the hypothesis (2.26) that

$$\sum_{j=0}^{k}|a_j|^q \leq \sup_{n\in\mathbb{N}}\sum_{j=0}^{\infty}|f_{nj}|^q < \infty$$

for all $k \in \mathbb{N}$. Hence, $(a_j) \in \ell_q$ which gives with Hölder's inequality that $(a_k y_k) \in \ell_1$ for all $y = (y_k) \in \ell_p$. For any given $\varepsilon > 0$, choose a fixed $k_0 \in \mathbb{N}$ such that

$$\left(\sum_{k=k_0+1}^{\infty} |y_k|^p \right)^{1/p} < \frac{\varepsilon}{4M^{1/q}} .$$

Then, there is some $n_0 \in \mathbb{N}$ by (2.28) such that

$$\left| \sum_{k=0}^{k_0} (f_{nk} - a_k)\, y_k \right| < \frac{\varepsilon}{2}$$

for every $n \geq n_0$. Therefore, we obtain with the relation (2.27) that

$$\begin{aligned} \left| \sum_k a_{nk} x_k - \sum_k a_k y_k \right| &= \left| \sum_k (f_{nk} - a_k)\, y_k \right| \\ &\leq \left| \sum_{k=0}^{k_0} (f_{nk} - a_k)\, y_k \right| + \left| \sum_{k=k_0+1}^{\infty} (f_{nk} - a_k)\, y_k \right| \\ &< \frac{\varepsilon}{2} + \left[\sum_{k=k_0+1}^{\infty} \left(|f_{nk}| + |a_k| \right)^q \right]^{1/q} \left(\sum_{k=k_0+1}^{\infty} |y_k|^p \right)^{1/p} \\ &< \frac{\varepsilon}{2} + 2M^{1/q} \frac{\varepsilon}{4M^{1/q}} = \varepsilon \end{aligned}$$

for all sufficiently large n. Hence, $Ax \in c$ which is what we wished to prove.

Conversely, suppose that $A = (a_{nk}) \in (\widetilde{\ell_p} : c)$ and take $x \in \widetilde{\ell_p}$. Since the inclusion relation $c \subset \ell_\infty$ holds, the necessity of the condition (2.26) follows from Theorem 2.2.27.

Now, consider the convergent sequence $u = (u_k) = \left\{ u_k^{(n)} \right\}_{k \in \mathbb{N}}$ defined by (2.21). Since A-transform of u exists and belongs to the space c by the hypothesis, one can see that $Au = (f_{nk})_{n \in \mathbb{N}} \in c$ for each fixed $k \in \mathbb{N}$ which shows the necessity of the condition (2.28).

This completes the proof. □

If the sequence space c is replaced by the space c_0, then Theorem 2.2.28 is reduced to

Corollary 2.2.29. *Let* $0 < p < \infty$. *Then,* $A = (a_{nk}) \in (\widetilde{\ell_p} : c_0)$ *if and only if*

(a) *In the case* $0 < p \leq 1$, *the condition (2.25) holds and (2.28) also holds with* $\alpha_k = 0$ *for all* $k \in \mathbb{N}$.

(b) *In the case* $1 < p < \infty$, *(2.26) holds and (2.28) also holds with* $\alpha_k = 0$ *for all* $k \in \mathbb{N}$.

Now, we can give our basic lemma which is useful for deriving the characterizations of the certain matrix classes via Theorems 2.2.27, 2.2.28 and Corollary 2.2.29.

Lemma 2.2.30. [25, Lemma 5.3] *Let λ and μ be any two sequence spaces, A be an infinite matrix and B a triangle matrix. Then, $A \in (\lambda : \mu_B)$ if and only if $BA \in (\lambda : \mu)$.*

It is trivial that Lemma 2.2.30 has several consequences related to the characterization of some new classes of matrix transformations. Indeed, combining Lemma 2.2.30 with Theorems 2.2.27, 2.2.28 and Corollary 2.2.29, one can easily derive the following results:

Corollary 2.2.31. *Let $A = (a_{nk})$ be an infinite matrix and define the matrix $C = (c_{nk})$ by*

$$c_{nk} = \sum_{j=0}^{n} \binom{n}{j} (1-r)^{n-j} r^j a_{jk}$$

for all $k, n \in \mathbb{N}$. Then, the necessary and sufficient conditions in order for A to belong to any one of the classes $(\widetilde{\ell_p} : e^r_\infty)$, $(\widetilde{\ell_p} : e^r_c)$ and $(\widetilde{\ell_p} : e^r_0)$ are obtained from the respective ones in Theorems 2.2.27, 2.2.28 and Corollary 2.2.29 by replacing the elements of the matrix A by those of the matrix C, where $0 < r < 1$.

Corollary 2.2.32. *Let $A = (a_{nk})$ be an infinite matrix and $t = (t_k)$ be a sequence of positive numbers, and define the matrix $C = (c_{nk})$ by*

$$c_{nk} = \frac{1}{T_n} \sum_{j=0}^{n} t_j a_{jk}$$

for all $k, n \in \mathbb{N}$, where $T_n = \sum_{k=0}^{n} t_k$ for all $n \in \mathbb{N}$. Then, the necessary and sufficient conditions in order for A to belong to any one of the classes $(\widetilde{\ell_p} : r^t_\infty)$, $(\widetilde{\ell_p} : r^t_c)$ and $(\widetilde{\ell_p} : r^t_0)$ are obtained from the respective ones in Theorems 2.2.27, 2.2.28 and Corollary 2.2.29 by replacing the elements of the matrix A by those of the matrix C.

Since the spaces r^t_∞, r^t_c and r^t_0 are reduced in the case $t = e$ to the Cesàro sequence spaces X_∞ and $\widetilde{c}$, $\widetilde{c}_0$ of non-absolute type, Corollary 3.2.22 also includes the characterization of the classes $(\widetilde{\ell_p} : X_\infty)$, $(\widetilde{\ell_p} : \widetilde{c})$ and $(\widetilde{\ell_p} : \widetilde{c}_0)$, as a special case.

Corollary 2.2.33. *Let $A = (a_{nk})$ be an infinite matrix and define the matrices $C = (c_{nk})$ and $D = (d_{nk})$ by $c_{nk} = a_{nk} - a_{n+1,k}$ and $d_{nk} = a_{nk} - a_{n-1,k}$ for all $k, n \in \mathbb{N}$. Then, the necessary and sufficient conditions in order for A to belong to any one of the classes $(\widetilde{\ell_p} : \ell_\infty(\Delta))$, $(\widetilde{\ell_p} : c(\Delta))$, $(\widetilde{\ell_p} : c_0(\Delta))$ and $(\widetilde{\ell_p} : bv_\infty)$ are obtained from the respective ones in Theorems 2.2.27, 2.2.28* *and Corollary 2.2.29 by replacing the elements of the matrix A by those of the matrices C and D, where $\ell_\infty(\Delta)$ denotes the difference space of all bounded sequences and were introduced by* Kızmaz [70].

Corollary 2.2.34. *Let $A = (a_{nk})$ be an infinite matrix and define the matrices $C = (c_{nk})$ and $D = (d_{nk})$ by $c_{nk} = \sum_{j=0}^{n} (1 + r^k)a_{jk}/(j + 1)$ and $d_{nk} = c_{nk} - c_{n-1,k}$ for all $k, n \in \mathbb{N}$. Then, the necessary and sufficient conditions in order for A to belong to any one of the classes $(\widetilde{\ell_p} : a_\infty^r)$, $(\widetilde{\ell_p} : a_0^r)$ and $(\widetilde{\ell_p} : a_c^r(\nabla))$, $(\widetilde{\ell_p} : a_0^r(\nabla))$ are obtained from the respective ones in Theorems 2.2.27, 2.2.28 and Corollary 2.2.29 by replacing the elements of the matrix A by those of the matrices C and D.*

Corollary 2.2.35. *Let $A = (a_{nk})$ be an infinite matrix and define the matrix $C = (c_{nk})$ by $c_{nk} = \sum_{j=0}^{n} a_{jk}$ for all $k, n \in \mathbb{N}$. Then, the necessary and sufficient conditions in order for A to belong to any one of the classes $(\widetilde{\ell_p} : bs)$, $(\widetilde{\ell_p} : cs)$ and $(\widetilde{\ell_p} : cs_0)$ are obtained from the respective ones in Theorems 2.2.27, 2.2.28 and Corollary 2.2.29 by replacing the elements of the matrix A by those of the matrix C.*

2.3 Spaces of Fibonacci Difference Sequences

Let $0 < p < 1$. In this section, following Başarır et al. [32], we introduce the sequence spaces $\ell_p(\widehat{F})$, $c_0(\widehat{F})$ and $c(\widehat{F})$ as the domain of the band matrix $\widehat{F}$ defined by the Fibonacci sequence in the classical sequence spaces ℓ_p, c_0 and c, respectively. Also, we give some inclusion relations and construct the bases of the spaces $c_0(\widehat{F})$ and $c(\widehat{F})$. Finally, we compute the alpha-, beta-, gamma-duals of these spaces and characterize the classes $(\ell_p(\widehat{F}) : \ell_\infty)$, $(\ell_p(\widehat{F}) : c)$ and $(\ell_p(\widehat{F}) : c_0)$ of infinite matrices.

The sequence (f_n) of Fibonacci numbers is defined by the linear recurrence equalities

$$f_0 = f_1 = 1 \quad \text{and} \quad f_n = f_{n-1} + f_{n-2} \quad \text{with} \quad n \geq 2.$$

Fibonacci numbers have many interesting properties and applications in arts, sciences and architecture. For example, the ratio sequences of Fibonacci numbers converge to the golden ratio which is important in sciences and arts. Some basic properties of sequences of Fibonacci numbers are also given below (cf. Koshy [74]):

$$\lim_{n\to\infty} \frac{f_{n+1}}{f_n} = \frac{1+\sqrt{5}}{2} = \varphi \text{ (Golden ratio)}, \tag{2.1}$$

$$\sum_{k=0}^{n} f_k = f_{n+2} - 1 \quad \text{for each} \quad n \in \mathbb{N},$$
$$\sum_k \frac{1}{f_k} \quad \text{converges},$$
$$f_{n-1}f_{n+1} - f_n^2 = (-1)^{n+1} \quad \text{for all } n \geq 1 \text{ (Cassini formula)}.$$

One can easily derive by substituting f_{n+1} in Cassini's formula that $f_{n-1}^2 + f_n f_{n-1} - f_n^2 = (-1)^{n+1}$.

Now, let $A = (a_{nk})$ be an infinite matrix and $\mathcal{F}$ be the collection of all finite subsets of $\mathbb{N}$. We list the following conditions:

$$\sup_{n\in\mathbb{N}} \sum_k |a_{nk}| < \infty \tag{2.2}$$
$$\lim_{n\to\infty} a_{nk} = 0 \quad \text{for each} \quad k \in \mathbb{N} \tag{2.3}$$
$$\exists \alpha_k \in \mathbb{R} \ni \lim_{n\to\infty} a_{nk} = \alpha_k \quad \text{for each} \quad k \in \mathbb{N} \tag{2.4}$$
$$\lim_{n\to\infty} \sum_k a_{nk} = 0 \tag{2.5}$$
$$\exists \alpha \in \mathbb{R} \ni \lim_{n\to\infty} \sum_k a_{nk} = \alpha \tag{2.6}$$
$$\sup_{K\in\mathcal{F}} \sum_n \left| \sum_{k\in K} a_{nk} \right| < \infty \tag{2.7}$$
$$\sup_{k,n\in\mathbb{N}} |a_{nk}| < \infty \tag{2.8}$$

Now, we can give the following lemma on the characterization of the classes of the matrix transformations between some classical sequence spaces.

Lemma 2.3.1. *The following statements hold* (see Stieglitz and Tietz [127])*:*

(a) $A = (a_{nk}) \in (c_0 : c_0)$ *if and only if (2.2) and (2.3) hold.*

(b) $A = (a_{nk}) \in (c_0 : c)$ *if and only if (2.2) and (2.4) hold.*

(c) $A = (a_{nk}) \in (c : c_0)$ *if and only if (2.2), (2.3) and (2.5) hold.*

(d) $A = (a_{nk}) \in (c : c)$ *if and only if (2.2), (2.4) and (2.6) hold.*

(e) $A = (a_{nk}) \in (c_0 : \ell_\infty) = (c : \ell_\infty)$ *if and only if the condition (2.2) holds.*

(f) $A = (a_{nk}) \in (c_0 : \ell_1) = (c : \ell_1)$ *if and only if the condition (2.7) holds.*

(g) $A = (a_{nk}) \in (\ell_p : c)$ *if and only if (2.4) and (2.8) hold, where* $0 < p < 1$.

(h) $A = (a_{nk}) \in (\ell_p : \ell_\infty)$ *if and only if the condition (2.8) holds, where* $0 < p < 1$.

2.3.1 The Fibonacci Difference Spaces of Absolutely p-Summable, Null and Convergent Sequences

In this subsection, we define the spaces $\ell_p(\widehat{F})$, $c_0(\widehat{F})$ and $c(\widehat{F})$ of Fibonacci absolutely p-summable, Fibonacci null and Fibonacci convergent sequences, where $0 < p < 1$. Also, we present some inclusion theorems and construct the Schauder bases for the spaces $\ell_p(\widehat{F})$, $c_0(\widehat{F})$ and $c(\widehat{F})$.

Recently, Kara [65] has defined the sequence space $\ell_p(\widehat{F})$ as follows:

$$\ell_p(\widehat{F}) = \left\{x \in \omega : \widehat{F}x \in \ell_p\right\}, \quad (1 \leq p \leq \infty),$$

where $\widehat{F} = (\widehat{f}_{nk})$ is the double band matrix defined by the sequence (f_n) of Fibonacci numbers as follows

$$\widehat{f}_{nk} = \begin{cases} -\frac{f_{n+1}}{f_n} & , \quad k = n-1, \\ \frac{f_n}{f_{n+1}} & , \quad k = n, \\ 0 & , \quad 0 \leq k < n-1 \text{ or } k > n \end{cases}$$

for all $k, n \in \mathbb{N}$. Also, in [66], Kara et al. have characterized some classes of compact operators on the spaces $\ell_p(\widehat{F})$ and $\ell_\infty(\widehat{F})$, where $1 \leq p < \infty$.

One can derive by a straightforward calculation that the inverse $\widehat{F}^{-1} = (g_{nk})$ of the Fibonacci matrix $\widehat{F}$ is given by

$$g_{nk} = \begin{cases} \frac{f_{n+1}^2}{f_k f_{k+1}} & , \quad 0 \leq k \leq n, \\ 0 & , \quad k > n \end{cases}$$

for all $k, n \in \mathbb{N}$.

Now, we introduce the Fibonacci difference sequence spaces $\ell_p(\widehat{F})$, $c_0(\widehat{F})$ and $c(\widehat{F})$ as the set of all sequences whose $\widehat{F}$-transforms are in the spaces ℓ_p, c_0 and c, respectively, i.e.,

$$\begin{aligned} \ell_p(\widehat{F}) &:= \left\{(x_n) \in \omega : \sum_n \left|\frac{f_n}{f_{n+1}}x_n - \frac{f_{n+1}}{f_n}x_{n-1}\right|^p < \infty\right\}, \ (0 < p < 1), \\ c_0(\widehat{F}) &:= \left\{(x_n) \in \omega : \lim_{n\to\infty}\left(\frac{f_n}{f_{n+1}}x_n - \frac{f_{n+1}}{f_n}x_{n-1}\right) = 0\right\}, \\ c(\widehat{F}) &:= \left\{(x_n) \in \omega : \exists l \in \mathbb{R} \ni \lim_{n\to\infty}\left(\frac{f_n}{f_{n+1}}x_n - \frac{f_{n+1}}{f_n}x_{n-1}\right) = l\right\}. \end{aligned}$$

With the notation of (4.4), the spaces $\ell_p(\widehat{F})$, $c_0(\widehat{F})$ and $c(\widehat{F})$ can be redefined as follows:

$$\ell_p(\widehat{F}) = (\ell_p)_{\widehat{F}}, \quad c_0(\widehat{F}) = (c_0)_{\widehat{F}} \quad \text{and} \quad c(\widehat{F}) = c_{\widehat{F}}.$$

Here and after, we assume unless stated otherwise in this subsection that $0 < p < 1$ and all the terms with negative subscript are equal to zero.

Define the sequence $y = (y_k)$ by the $\widehat{F}$-transform of a sequence $x = (x_k)$, i.e.,

$$y_k = (\widehat{F}x)_k = \frac{f_k}{f_{k+1}}x_k - \frac{f_{k+1}}{f_k}x_{k-1} \tag{2.9}$$

for all $k \in \mathbb{N}$. Therefore, one can derive by a straightforward calculation that

$$x_k = \sum_{j=0}^{k} \frac{f_{k+1}^2}{f_j f_{j+1}} y_j \quad \text{for all} \quad k \in \mathbb{N}. \tag{2.10}$$

Throughout the text, we suppose that the sequences $x = (x_k)$ and $y = (y_k)$ are connected with the relation (2.9).

Theorem 2.3.2. *The following statements hold:*

(i) The sets $\ell_p(\widehat{F})$, $c_0(\widehat{F})$ and $c(\widehat{F})$ are the linear spaces with the coordinatewise addition and scalar multiplication.

(ii) $\ell_p(\widehat{F})$ is a complete p-normed space with the p-norm $\|x\|_p = \sum_n \left|\widehat{F}_n(x)\right|^p$.

(iii) $c_0(\widehat{F})$ and $c(\widehat{F})$ are the BK-spaces with the norm $\|x\|_{c_0(\widehat{F})} = \|x\|_{c(\widehat{F})} = \|\widehat{F}x\|_\infty$.

Proof. This is a routine verification and so we omit details. □

Remark 2.3.3. *One can easily check that the absolute property does not hold on the spaces $\ell_p(\widehat{F})$, $c_0(\widehat{F})$ and $c(\widehat{F})$, that is $\|x\|_p \neq \||x|\|_p$, $\|x\|_{c_0(\widehat{F})} \neq \||x|\|_{c_0(\widehat{F})}$ and $\|x\|_{c(\widehat{F})} \neq \||x|\|_{c(\widehat{F})}$ for at least one sequence in the spaces $\ell_p(\widehat{F})$, $c_0(\widehat{F})$ and $c(\widehat{F})$, and this means that $\ell_p(\widehat{F})$, $c_0(\widehat{F})$ and $c(\widehat{F})$ are the sequence spaces of non-absolute type, where $|x| = (|x_k|)$.*

Let λ denotes any of the spaces ℓ_p, c_0 and c. With the notation of (2.9), since the transformation $T : \lambda(\widehat{F}) \to \lambda$ defined by $x \mapsto y = Tx = \widehat{F}x$ is a norm preserving linear bijection, we have the following:

Corollary 2.3.4. *The Fibonacci difference sequence spaces $\ell_p(\widehat{F})$, $c_0(\widehat{F})$ and $c(\widehat{F})$ of non-absolute type are linearly p-norm/norm isomorphic to the spaces ℓ_p, c_0 and c, respectively, i.e., $\ell_p(\widehat{F}) \cong \ell_p$, $c_0(\widehat{F}) \cong c_0$ and $c(\widehat{F}) \cong c$.*

Now, we give some inclusion relations concerning the spaces $c_0(\widehat{F})$ and $c(\widehat{F})$.

Theorem 2.3.5. *The inclusion $c_0(\widehat{F}) \subset c(\widehat{F})$ strictly holds.*

Proof. It is clear that the inclusion $c_0(\widehat{F}) \subset c(\widehat{F})$ holds. Further, to show the strictness of the inclusion, consider the sequence $x = (x_k) = \left(\sum_{j=0}^{k} f_{k+1}^2/f_j^2\right)$. Then, we obtain by (2.9) for all $k \in \mathbb{N}$ that

$$(\widehat{F}x)_k = \frac{f_k}{f_{k+1}} \sum_{j=0}^{k} \frac{f_{k+1}^2}{f_j^2} - \frac{f_{k+1}}{f_k} \sum_{j=0}^{k-1} \frac{f_k^2}{f_j^2} = \frac{f_{k+1}}{f_k}$$

which tends to φ, as $k \to \infty$ by (2.1). That is to say that $\widehat{F}x \in c \backslash c_0$. Thus, the sequence x is in $c(\widehat{F})$ but is not in $c_0(\widehat{F})$. Hence, the inclusion $c_0(\widehat{F}) \subset c(\widehat{F})$ is strict. □

Theorem 2.3.6. *The space ℓ_∞ does not include the spaces $c_0(\widehat{F})$ and $c(\widehat{F})$.*

Proof. Let us consider the sequence $x = (x_k) = (f_{k+1}^2)$. Since $f_{k+1}^2 \to \infty$, as $k \to \infty$ and $\widehat{F}x = e^{(0)} = (1, 0, 0, \ldots)$, the sequence x is in the space $c_0(\widehat{F})$ but is not in the space ℓ_∞. This shows that the space ℓ_∞ does not include both the space $c_0(\widehat{F})$ and the space $c(\widehat{F})$, as desired. □

Theorem 2.3.7. *The inclusions $c_0 \subset c_0(\widehat{F})$ and $c \subset c(\widehat{F})$ strictly hold.*

Proof. Let $\lambda = c_0$ or c. Since the matrix $\widehat{F} = (\widehat{f}_{nk})$ satisfies the conditions

$$\sup_{n \in \mathbb{N}} \sum_k \left|\widehat{f}_{nk}\right| = \sup_{n \in \mathbb{N}} \left(\frac{f_n}{f_{n+1}} + \frac{f_{n+1}}{f_n}\right) = 2 + \frac{1}{2} = \frac{5}{2},$$

$$\lim_{n \to \infty} \widehat{f}_{nk} = 0,$$

$$\lim_{n \to \infty} \sum_k \widehat{f}_{nk} = \lim_{n \to \infty} \left(\frac{f_n}{f_{n+1}} - \frac{f_{n+1}}{f_n}\right) = \frac{1}{\varphi} - \varphi$$

we conclude by Parts (a) and (d) of Lemma 2.3.1 that $\widehat{F} \in (\lambda : \lambda)$. This leads to the fact that $\widehat{F}x \in \lambda$ for any $x \in \lambda$. Thus, $x \in \lambda(\widehat{F})$. This shows that $\lambda \subset \lambda(\widehat{F})$.

Now, let $x = (x_k) = (f_{k+1}^2)$. Then, it is clear that $x \in \lambda(\widehat{F}) \setminus \lambda$. This says that the inclusion $\lambda \subset \lambda(\widehat{F})$ is strict. □

Theorem 2.3.8. *The spaces $c_0(\widehat{F})$ and $c(\widehat{F})$ are not solid.*

Proof. Consider the sequences $u = (u_k)$ and $v = (v_k)$ defined by $u_k = f_{k+1}^2$ and $v_k = (-1)^{k+1}$ for all $k \in \mathbb{N}$. Then, it is clear that $u \in c_0(\widehat{F})$ and $v \in \ell_\infty$. Nevertheless $uv = \{(-1)^{k+1} f_{k+1}^2\}$ does not in the space $c_0(\widehat{F})$, since

$$(\widehat{F}uv)_k = \frac{f_k}{f_{k+1}}(-1)^{k+1} f_{k+1}^2 - \frac{f_{k+1}}{f_k}(-1)^k f_k^2 = 2(-1)^{k+1} f_k f_{k+1}$$

for all $k \in \mathbb{N}$. This shows that the multiplication $\ell_\infty c_0(\widehat{F})$ of the spaces ℓ_∞ and $c_0(\widehat{F})$ is not a subset of $c_0(\widehat{F})$. Hence, the space $c_0(\widehat{F})$ is not solid.

It is clear here that if the space $c_0(\widehat{F})$ is replaced by the space $c(\widehat{F})$, then we obtain the fact that $c(\widehat{F})$ is not solid.

This completes the proof. □

It is known from Theorem 2.3 of Jarrah and Malkowsky [60] that if T is a triangle then the domain λ_T of T in a normed sequence space λ has a basis if and only if λ has a basis. As a direct consequence of this fact, we have:

Corollary 2.3.9. *Define the sequences* $c^{(-1)} = \{c_k^{(-1)}\}_{k\in\mathbb{N}}$ *and* $c^{(n)} = \{c_k^{(n)}\}_{k\in\mathbb{N}}$ *for every fixed* $n \in \mathbb{N}$ *by*

$$c_k^{(-1)} = \sum_{j=0}^{k} \frac{f_{k+1}^2}{f_j f_{j+1}} \quad \text{and} \quad c_k^{(n)} = \begin{cases} 0 & , \quad 0 \le k \le n-1, \\ \dfrac{f_{k+1}^2}{f_n f_{n+1}} & , \quad k \ge n. \end{cases}$$

Then, the following statements hold:

(a) The sequence $\{c^{(n)}\}_{n=0}^{\infty}$ *is a basis for the spaces* $\ell_p(\widehat{F})$ *and* $c_0(\widehat{F})$, *and every sequence* $x \in c_0(\widehat{F})$ *or in the space* $\ell_p(\widehat{F})$ *has a unique representation of the form* $x = \sum_n \widehat{F}_n(x) c^{(n)}$.

(b) The sequence $\{c^{(n)}\}_{n=-1}^{\infty}$ *is a basis for the space* $c(\widehat{F})$ *and every sequence* $z = (z_n) \in c(\widehat{F})$ *has a unique representation of the form* $z = lc^{(-1)} + \sum_n \left[\widehat{F}_n(z) - l\right] c^{(n)}$, *where* $l = \lim_{n\to\infty} \widehat{F}_n(z)$.

2.3.2 The Alpha-, Beta- and Gamma-duals of the Spaces $\ell_p(\widehat{F})$, $c_0(\widehat{F})$ and $c(\widehat{F})$, and Some Matrix Transformations

In this subsection, we determine the alpha, beta and gamma duals of the spaces $\ell_p(\widehat{F})$, $c_0(\widehat{F})$ and $c(\widehat{F})$, and characterize the classes of infinite matrices from the spaces $\ell_p(\widehat{F})$ to the spaces ℓ_∞, c and c_0.

Theorem 2.3.10. *The alpha dual of the spaces* $c_0(\widehat{F})$ *and* $c(\widehat{F})$ *is the set*

$$d_1 := \left\{ a = (a_k) \in \omega : \sup_{K\in\mathcal{F}} \sum_n \left| \sum_{k\in K} b_{nk} \right| < \infty \right\},$$

where the matrix $B = (b_{nk})$ *is defined by*

$$b_{nk} = \begin{cases} \dfrac{f_{n+1}^2}{f_k f_{k+1}} a_n & , \quad 0 \le k \le n, \\ 0 & , \quad k > n \end{cases}$$

for all $k, n \in \mathbb{N}$.

Proof. Let $a = (a_n) \in \omega$. Consider the equality

$$a_n x_n = \sum_{k=0}^{n} \frac{f_{n+1}^2}{f_k f_{k+1}} a_n y_k = \sum_{k=0}^{n} b_{nk} y_k = (By)_n. \tag{2.11}$$

By (2.11), we obtain that $ax = (a_n x_n) \in \ell_1$ whenever $x = (x_k) \in c_0(\widehat{F})$ or $c(\widehat{F})$ if and only if $By \in \ell_1$ whenever $y = (y_k) \in c_0$ or c. That is, the sequence $a = (a_n)$ is in the alpha dual of the space $c_0(\widehat{F})$ or $c(\widehat{F})$ if and only if $B \in (c_0 : \ell_1) = (c : \ell_1)$. By taking B instead of A in part (f) of Lemma 2.3.1, we obtain that $a \in \left[c_0(\widehat{F})\right]^\alpha = \left[c(\widehat{F})\right]^\alpha$ if and only if

$$\sup_{K \in \mathcal{F}} \sum_n \left| \sum_{k \in K} b_{nk} \right| < \infty$$

which means that $\left[c_0(\widehat{F})\right]^\alpha = \left[c(\widehat{F})\right]^\alpha = d_1$. □

Theorem 2.3.11. *Define the sets d_2, d_3, d_4 and d_5, as follows:*

$$d_2 := \left\{ a = (a_k) \in \omega : \sup_{n \in \mathbb{N}} \sum_{k=0}^{n} \left| \sum_{j=k}^{n} \frac{f_{j+1}^2}{f_k f_{k+1}} a_j \right| < \infty \right\},$$

$$d_3 := \left\{ a = (a_k) \in \omega : \lim_{n \to \infty} \sum_{j=k}^{n} \frac{f_{j+1}^2}{f_k f_{k+1}} a_j \text{ exists for each } k \in \mathbb{N} \right\},$$

$$d_4 := \left\{ a = (a_k) \in \omega : \lim_{n \to \infty} \sum_{k=0}^{n} \sum_{j=k}^{n} \frac{f_{j+1}^2}{f_k f_{k+1}} a_j \text{ exists} \right\},$$

$$d_5 := \left\{ a = (a_k) \in \omega : \sup_{k,n \in \mathbb{N}} \left| \sum_{j=k}^{n} \frac{f_{j+1}^2}{f_k f_{k+1}} a_j \right| < \infty \right\}.$$

Then, the following statements hold:

(a) $\left[c_0(\widehat{F})\right]^\alpha = \left[c(\widehat{F})\right]^\alpha = d_1$.

(b) $\left[c_0(\widehat{F})\right]^\beta = d_2 \cap d_3$.

(c) $\left[c(\widehat{F})\right]^\beta = d_2 \cap d_3 \cap d_4$.

(d) $\left[\ell_p(\widehat{F})\right]^\beta = d_3 \cap d_5$.

(e) $\left[c_0(\widehat{F})\right]^\gamma = \left[c(\widehat{F})\right]^\gamma = d_2$.

(f) $\left[\ell_p(\widehat{F})\right]^\gamma = d_5$.

Proof. Consider the equality

$$\sum_{k=0}^{n} a_k x_k = \sum_{k=0}^{n} a_k \left(\sum_{i=0}^{k} \frac{f_{k+1}^2}{f_i f_{i+1}} y_i \right) = \sum_{k=0}^{n} \left(\sum_{i=k}^{n} \frac{f_{i+1}^2}{f_k f_{k+1}} a_i \right) y_k = (Cy)_n, \tag{2.12}$$

where $C = (c_{nk})$ defined by

$$c_{nk} = \begin{cases} \sum\limits_{i=k}^{n} \dfrac{f_{i+1}^2}{f_k f_{k+1}} a_i & , \quad 0 \leq k \leq n \\ 0 & , \quad k > n \end{cases}$$

for all $k, n \in \mathbb{N}$. Then, we observe by (3.11) that $ax = (a_n x_n) \in cs$ whenever $x \in c_0(\widehat{F})$ if and only if $Cy \in c$ whenever $y = (y_k) \in c_0$ which means that $a \in \left[c_0(\widehat{F})\right]^{\beta}$ if and only if $C \in (c_0 : c)$. By using Part (b) of Lemma 2.3.1, we derive that

$$\sup_{n \in \mathbb{N}} \sum_{k} \left| \sum_{i=k}^{n} \frac{f_{i+1}^2}{f_k f_{k+1}} a_i \right| < \infty,$$

$$\lim_{n \to \infty} \sum_{i=k}^{n} \frac{f_{i+1}^2}{f_k f_{k+1}} a_i \text{ exists for each } k \in \mathbb{N}.$$

Hence, we conclude that $\left[c_0(\widehat{F})\right]^{\beta} = d_2 \cap d_3$.

In a similar way, one can easily show the facts about the beta duals of the spaces $c(\widehat{F})$, $\ell_p(\widehat{F})$ and gamma duals of the spaces $c_0(\widehat{F})$, $c(\widehat{F})$, $\ell_p(\widehat{F})$. □

Now, we give the theorems characterizing the classes $(\ell_p(\widehat{F}) : \ell_\infty)$, $(\ell_p(\widehat{F}) : c)$ and $(\ell_p(\widehat{F}) : c_0)$ of infinite matrices.

Theorem 2.3.12. *$A = (a_{nk}) \in (\ell_p(\widehat{F}) : \ell_\infty)$ if and only if*

$$\sup_{k,n \in \mathbb{N}} \left| \sum_{j=k}^{\infty} \frac{f_{j+1}^2}{f_k f_{k+1}} a_{nj} \right| < \infty. \tag{2.13}$$

Proof. $A = (a_{nk}) \in (\ell_p(\widehat{F}) : \ell_\infty)$ and $x = (x_k) \in \ell_p(\widehat{F})$. Then, Ax exists and belongs to the space ℓ_∞. Therefore, one can conclude for $x^{(k)} = \left\{x_j^{(k)}\right\} \in \ell_p(\widehat{F})$ defined by

$$x_j^{(k)} = \begin{cases} \dfrac{f_{j+1}^2}{f_k f_{k+1}} & , \quad j \geq k \\ 0 & , \quad 0 \leq j \leq k-1 \end{cases} \tag{2.14}$$

for all $j,k\in\mathbb{N}$ that $Ax^{(k)}=\left(\sum_{j=k}^{\infty}f_{j+1}^2a_{nj}/f_kf_{k+1}\right)\in\ell_\infty$ for each $k\in\mathbb{N}$. Hence, the condition (2.13) is necessary.

Conversely, suppose that (2.13) holds and take any $x=(x_k)\in\ell_p(\widehat{F})$. Then, $A_n\in[\ell_p(\widehat{F})]^\beta$ for each $n\in\mathbb{N}$ which leads to the existence of Ax. Let $n\in\mathbb{N}$ be fixed. Consider the following relation derived from the mth partial sum of the series $\sum_k a_{nk}x_k$ with (2.10):

$$\sum_{k=0}^{m}a_{nk}x_k=\sum_{k=0}^{m}\sum_{j=k}^{m}\frac{f_{j+1}^2}{f_kf_{k+1}}a_{nj}y_k \tag{2.15}$$

for all $m,n\in\mathbb{N}$. Then, by letting $m\to\infty$ in (2.15) we have

$$(Ax)_n=\sum_k a_{nk}x_k=\sum_k e_{nk}y_k=(Ey)_n \tag{2.16}$$

for all $n\in\mathbb{N}$, where the matrix $E=(e_{nk})$ is defined by

$$e_{nk}=\sum_{j=k}^{\infty}\frac{f_{j+1}^2}{f_kf_{k+1}}a_{nj} \tag{2.17}$$

for all $k,n\in\mathbb{N}$. Therefore, since

$$\begin{aligned}\left|\sum_k a_{nk}x_k\right|^p &= \left|\sum_k\sum_{j=k}^{\infty}\frac{f_{j+1}^2}{f_kf_{k+1}}a_{nj}y_k\right|^p\\ &\le \left(\sum_k\left|\sum_{j=k}^{\infty}\frac{f_{j+1}^2}{f_kf_{k+1}}a_{nj}\right||y_k|\right)^p\\ &\le \left(\sup_{k\in\mathbb{N}}\left|\sum_{j=k}^{\infty}\frac{f_{j+1}^2}{f_kf_{k+1}}a_{nj}\right|\right)^p\left(\sum_k|y_k|\right)^p\\ &\le \left(\sup_{k\in\mathbb{N}}\left|\sum_{j=k}^{\infty}\frac{f_{j+1}^2}{f_kf_{k+1}}a_{nj}\right|\right)^p\sum_k|y_k|^p\end{aligned}$$

we obtain by taking supremum over $n\in\mathbb{N}$ that

$$\|Ax\|_\infty=\sup_{n\in\mathbb{N}}\left|\sum_k a_{nk}x_k\right|\le(\|y\|_p)^{1/p}\sup_{k,n\in\mathbb{N}}\left|\sum_{j=k}^{\infty}\frac{f_{j+1}^2}{f_kf_{k+1}}a_{nj}\right|<\infty.$$

That is, $Ax\in\ell_\infty$, as desired.

This completes the proof. □

Theorem 2.3.13. *$A = (a_{nk}) \in (\ell_p(\widehat{F}) : c)$ if and only if (2.13) holds, and*

$$\exists \alpha_k \in \mathbb{R} \text{ such that } \lim_{n\to\infty} \sum_{j=k}^{\infty} \frac{f_{j+1}^2}{f_k f_{k+1}} a_{nj} = \alpha_k \tag{2.18}$$

for each $k \in \mathbb{N}$.

Proof. Let $A = (a_{nk}) \in (\ell_p(\widehat{F}) : c)$. Then, Ax exists and is in the space c for all $x = (x_k) \in \ell_p(\widehat{F})$. Since the inclusion $c \subset \ell_\infty$ holds, the necessity of the condition (2.13) follows from Theorem 2.3.12. The necessity of the condition (2.18) is immediate by taking the sequence $x^{(k)} = \{x_j^{(k)}\} \in \ell_p(\widehat{F})$ defined by (2.14).

Conversely, suppose that the conditions (2.13) and (2.18) hold, and take any $x = (x_k) \in \ell_p(\widehat{F})$. Then, since $A_n \in \{\ell_p(\widehat{F})\}^\beta$ for each $n \in \mathbb{N}$, Ax exists. Then, by taking into account the relation (2.16) one can see that the conditions (2.13) and (2.18) correspond to (2.8) and (2.4) with e_{nk} instead of a_{nk}, respectively, where e_{nk} is given by (2.17). Hence, $Ey \in c$ which gives by (2.16) that $A \in (\ell_p(\widehat{F}) : c)$. □

If we replace the space c_0 with the space c, then Theorem 2.3.13 yields the following:

Corollary 2.3.14. *$A = (a_{nk}) \in (\ell_p(\widehat{F}) : c_0)$ if and only if (2.13) holds, and (2.18) also holds with $\alpha_k = 0$ for all $k \in \mathbb{N}$.*

Theorem 2.3.15. *Suppose that the elements of the infinite matrices $A = (a_{nk})$ and $H = (h_{nk})$ are connected with the relation*

$$h_{nk} = -\frac{f_{n+1}}{f_n} a_{n-1,k} + \frac{f_n}{f_{n+1}} a_{nk} \tag{2.19}$$

for all $k, n \in \mathbb{N}$ and μ be any given sequence space. Then, $A \in (\mu : \lambda(\widehat{F}))$ if and only if $H \in (\mu : \lambda)$; where λ denotes any of the classical sequence spaces ℓ_p, c_0 or c.

Proof. Let $z = (z_k) \in \mu$. Then, by taking into account the relation (2.19) one can easily derive the following equality

$$\sum_{k=0}^{m} h_{nk} z_k = \sum_{k=0}^{m} \left(-\frac{f_{n+1}}{f_n} a_{n-1,k} + \frac{f_n}{f_{n+1}} a_{nk} \right) z_k \quad \text{for all } m, n \in \mathbb{N}$$

which yields as $m \to \infty$ that $(Hz)_n = (\widehat{F}Az)_n$. Therefore, we conclude that $Az \in \lambda(\widehat{F})$ whenever $z \in \mu$ if and only if $Hz \in \lambda$ whenever $z \in \mu$.

This step completes the proof. □

It is trivial that combining Theorems 2.3.12, 2.3.13 and Corollary 2.3.14 with Theorem 2.3.15, one can derive the following results:

Corollary 2.3.16. *Let $A = (a_{nk})$ be an infinite matrix and $a(n,k) = \sum_{j=0}^{n} a_{jk}$ for all $k, n \in \mathbb{N}$. Then, the following statements hold:*

(a) $A = (a_{nk}) \in (\ell_p(\widehat{F}), bs)$ if and only if (2.13) holds with $a(n,k)$ instead of a_{nk}.

(b) $A = (a_{nk}) \in (\ell_p(\widehat{F}), cs)$ if and only if (2.13) and (2.18) hold with $a(n,k)$ instead of a_{nk}.

(c) $A = (a_{nk}) \in (\ell_p(\widehat{F}), cs_0)$ if and only if (2.13) and (2.18) hold with $a(n,k)$ instead of a_{nk} with $\alpha_k = 0$ for all $k \in \mathbb{N}$.

(d) $A = (a_{nk}) \in (\ell_p(\widehat{F}), \ell_\infty(\widehat{F}))$ if and only if (2.13) holds with h_{nk} instead of a_{nk}.

(e) $A = (a_{nk}) \in (\ell_p(\widehat{F}), c(\widehat{F}))$ if and only if (2.13) and (2.18) hold with h_{nk} instead of a_{nk}.

(f) $A = (a_{nk}) \in (\ell_p(\widehat{F}), c_0(\widehat{F}))$ if and only if (2.13) and (2.18) hold with h_{nk} instead of a_{nk} with $\alpha_k = 0$ for all $k \in \mathbb{N}$.

2.4 Conclusion

Lorentz [81] introduced the concept of almost convergence in 1948. Başar and Kirişçi [29] determined the beta- and gamma-duals of the spaces f, fs and $\widehat{f}$, and proved some basic results on the space f and characterized the class of matrix transformations from the space $\widehat{f}$ into any given sequence space, where fs denotes the space of almost convergent series.

Kayaduman and Şengönül have investigated the spaces $\overline{f}_0$ and $\overline{f}$ that consist of all sequences whose Cesàro mean of order one transforms are in the spaces f_0 and f in [69], respectively. Şengönül and Kayaduman [129] have defined the spaces $\widehat{f}_0$ and $\widehat{f}$ as the domain of Riesz mean in the sequence spaces f_0 and f. They also showed that the spaces $\widehat{f}_0$ and $\widehat{f}$ are linearly isomorphic to the sequence spaces f_0 and f, respectively. After computing the beta and gamma-duals of $\widehat{f}_0$ and $\widehat{f}$ they characterized the classes $(\widehat{f} : \mu)$ and $(\mu : \widehat{f})$ of infinite matrices and determined some core theorems related to the space $\widehat{f}$. Recently, Sönmez [126] worked the domain $f(B)$ of the triple band matrix $B(r, s, t)$ in the space f.

Candan [38] has established the sequence spaces $f(\widetilde{B})$ and $f_0(\widetilde{B})$ as the domain of the double sequential band matrix $\widetilde{B}(\widetilde{r}, \widetilde{s})$ in the sequence spaces f_0 and f. In this study, Candan has determined the beta- and gamma-duals of the spaces $f_0(\widetilde{B})$ and $f(\widetilde{B})$, and also gave some inclusion theorems related with the spaces $f_0(\widetilde{B})$ and $f(\widetilde{B})$. Finally, he has recently characterized the classes $(f(\widetilde{B}) : \mu)$ and $(\mu : f(\widetilde{B}))$ of infinite matrices.

Kirişci [72] has studied the domains $(f_0)_{E^r}$ and f_{E^r} of the Euler means of order r in the spaces f_0 and f, respectively. Yeşilkayagil and Başar [138] have presented the domains $A_\lambda(f_0)$ and $A_\lambda(f)$ of the matrix A_λ in the spaces f_0 and f, respectively, and established some inclusion relations dealing with sequence spaces. The alpha-, beta- and gamma-duals of the sequence spaces $A_\lambda(f_0)$ and $A_\lambda(f)$ have been computed and characterization of the classes $(A_\lambda(f) : \mu)$ and $(\mu : A_\lambda(f_0))$ of infinite matrices have been given.

In 1978, the domain of Nörlund matrix N^t in the classical sequence spaces ℓ_∞ and ℓ_p were introduced by Wang [133], where $1 \leq p < \infty$. In 1978, the domain of Cesàro matrix C of order one in the classical sequence spaces ℓ_∞ and ℓ_p were introduced by Ng and Lee [115], where $1 \leq p < \infty$. Following Ng and Lee [115], Şengönül and Başar [128] have studied the domain of Cesàro matrix C of order one in the classical sequence spaces c_0 and c. Following Şengönül and Başar [128], Tuğ and Başar [130] have worked on the domain of Nörlund matrix N^t in the classical sequence spaces c_0 and c.

Since in the special case $t = e$, the Nörlund mean N^t is reduced to the Cesàro mean C_1 of order one, the corresponding results are much more general than those given by Kayaduman and Şengönül, in [69].

Although the matrix transformations from the domain of certain triangles in the classical sequence spaces into the classical sequence spaces have been characterized, the matrix transformations from the domain of Nörlund matrix in the spaces of null and convergent sequences into some classical sequence spaces have been characterized.

In [22, 23], the domain of the triangle $\widetilde{B}$ in the space ℓ_p has recently been investigated for $1 \leq p < \infty$ and $0 < p \leq 1$, respectively. Quite recently, following Başar [22, 23], Başar and Braha have studied the domains of the matrix $\widetilde{B}$ in the classical sequence spaces ℓ_∞, c and c_0 in [27]. Furthermore, one may also investigate some geometric properties, for example Banach-Saks, weak Banach-Saks and fixed point properties of the sequence space $\widetilde{\ell_p}$, in the case $1 \leq p < \infty$.

Reviewing the literature about the domain of the infinite matrix A in the sequence spaces ℓ_p, c_0 and c, Table 2.1, below, may be useful.

Finally, we have introduced the sequence spaces $\ell_p(\widehat{F})$, $c_0(\widehat{F})$ and $c(\widehat{F})$ obtained as the domain of the double band matrix $\widehat{F}$ in the classical spaces ℓ_p, c_0 and c, respectively. We should state that although the double sequential band matrix $B(\widetilde{r}, \widetilde{s})$ can be reduced to the matrix $\widehat{F}$ in the case $\widetilde{r} = (r_n)$ and $\widetilde{s} = (s_n)$ with $r_n = f_n/f_{n+1}$ and $s_{n-1} = -f_{n+1}/f_n$ for all $n \in \mathbb{N}$, the main results concerning the spaces $c_0(\widehat{F})$ and $c(\widehat{F})$ are obtained independently from Candan [37]. It is worth mentioning here that in spite of the domain of the matrix $\widehat{F}$ in the space ℓ_p of absolutely p-summable sequences has been recently studied by Kara in [65] for $1 \leq p < \infty$, the results related to the space $\ell_p(\widehat{F})$ with $0 < p < 1$ are complementary of Kara [65].

TABLE 2.1: The domains of some triangle matrices in the spaces ℓ_p, c_0 and c.

A	λ	λ_A	refer to:
Δ	c_0 and c	$c_0(\Delta)$ and $c(\Delta)$	[70]
∇	ℓ_p, $(0 < p \leq 1)$	bv_p	[10]
$B(r,s,t)$	c_0 and c	$B(c_0)$ and $B(c)$	[125]
C	c_0 and c	$\widetilde{c_0}$ and $\widetilde{c}$	[128]
A^r	c_0 and c	a_0^r and a_c^r	[13]
Δ^2	c_0 and c	$c_0(\Delta^2)$ and $c(\Delta^2)$	[56]
$u\Delta^2$	c_0 and c	$c_0(u;\Delta^2)$ and $c(u;\Delta^2)$	[103]
Δ^m	c_0 and c	$c_0(\Delta^m)$ and $c(\Delta^m)$	[49, 45]
R^q	c_0 and c	$(\overline{N},q)_0$ and $(\overline{N},q)$	[92]
$\Delta^{(m)}$	c_0 and c	$c_0(\Delta^{(m)})$ and $c(\Delta^{(m)})$	[96]
$G(u,v)$	c_0 and c	$c_0(u,v)$ and $c(u,v)$	[7]
Λ	c_0 and c	c_0^λ and c^λ	[105]
$B(r,s)$	c_0 and c	$\widehat{c_0}$ and $\widehat{c}$	[73]
E^r	c_0 and c	e_0^r and e_c^r	[5]
A_λ	c_0 and c	$A_\lambda(c_0)$ and $A_\lambda(c)$	[36]
$B(\widetilde{r},\widetilde{s})$	c_0 and c	$\widetilde{c_0}$ and $\widetilde{c}$	[37]
N^t	c_0 and c	$c_0(N^t)$ and $c(N^t)$	[130]

Consider the matrix $\widetilde{B} = (\widetilde{b}_{nk})$ defined by (2.1) as the composition of the matrices E_1, C and ∇. We should record from now on that one can study the domain of the matrix $\widetilde{B}$ in the spaces f and f_0 of almost null and almost convergent sequences for introducing some new normed spaces, as a natural continuation of this chapter.

Chapter 3

Some Paranormed Spaces Derived by the Double Sequential Band Matrix

Abstract. *In the present chapter, we study the sequence spaces* $\ell_\infty(\widetilde{B},p)$, $c(\widetilde{B},p)$ *and* $c_0(\widetilde{B},p)$ *generated by the domains of double sequential band matrix* $B(\widetilde{r},\widetilde{s})$ *in the Maddox's spaces* $\ell_\infty(p)$, $c(p)$, $c_0(p)$ *and* $\ell(p)$, *and investigate their topological properties.*

Keywords. *Domain of an infinite matrix, Double sequential band matrix, Paranormed sequence spaces, Characterization of the matrix transformations.*

3.1 Domains of the Double Sequential Band Matrix in Some Maddox's Spaces

In this section, we emphasize the domains of the double sequential band matrix $B(\widetilde{r},\widetilde{s})$ in the Maddox's spaces $\ell_\infty(p)$, $c(p)$, $c_0(p)$ and $\ell(p)$.

Let $\widetilde{r}=(r_k)$ and $\widetilde{s}=(s_k)$ be the convergent sequences whose entries are either constants or distinct non-zero numbers. Define the double sequential band matrix $B(\widetilde{r},\widetilde{s})=\{b_{nk}(r_k,s_k)\}_{n,k=0}^{\infty}$ by

$$b_{nk}(r_k,s_k)=\begin{cases} r_k & , \quad k=n, \\ s_k & , \quad k=n-1, \\ 0 & , \quad \text{otherwise} \end{cases}$$

for all $k,n\in\mathbb{N}$. We should note that the double sequential band matrices were firstly used by Srivastava and Kumar [123, 124], Panigrahi and Srivastava [118] and Akhmedov and El-Shabrawy [2]. It is clear that ∇ can be obtained as a special case of $B(\widetilde{r},\widetilde{s})$ for $\widetilde{r}=e$ and $\widetilde{s}=-e$, and it is also trivial that $B(\widetilde{r},\widetilde{s})$ is reduced in the special case $\widetilde{r}=re$ and $\widetilde{s}=se$ to the generalized difference matrix $B(r,s)$. So, the results related to the domain of the matrix $B(\widetilde{r},\widetilde{s})$ are more general and comprehensive than the corresponding consequences of the domain of the matrices ∇ and $B(r,s)$.

Define the sequence $y=(y_k)$, which will be frequently used, by the $B(\widetilde{r},\widetilde{s})$-transform of a sequence $x=(x_k)$, that is,

$$y_k=\{B(\widetilde{r},\widetilde{s})x\}_k=r_kx_k+s_{k-1}x_{k-1} \quad \text{for all} \quad k\in\mathbb{N}. \tag{3.1}$$

Throughout Sections 3.1 and 3.2, we suppose that the terms of the sequences $x = (x_k)$ and $y = (y_k)$ are connected with the relation (3.1).

3.1.1 The Spaces $\ell_\infty(\widetilde{B},p)$, $c(\widetilde{B},p)$ and $c_0(\widetilde{B},p)$, and Their Topological Properties

In this subsection, following Özger and Başar [116, 117], we introduce the sequence spaces $\ell_\infty(\widetilde{B},p)$, $c(\widetilde{B},p)$ and $c_0(\widetilde{B},p)$ of non-absolute type derived as the domain of the double sequential band matrix $B(\widetilde{r},\widetilde{s})$ in the Maddox spaces $\ell_\infty(p)$, $c(p)$ and $c_0(p)$, respectively, and prove that these are the complete paranormed linear spaces.

The sequence spaces $\ell_\infty(\widetilde{B},p)$, $c(\widetilde{B},p)$ and $c_0(\widetilde{B},p)$ are defined as the set of all sequences whose $B(\widetilde{r},\widetilde{s})$-transforms are in the spaces $\ell_\infty(p)$, $c(p)$ and $c_0(p)$, respectively, that is,

$$\begin{aligned}
\ell_\infty(\widetilde{B},p) &:= \left\{x = (x_k) \in \omega : \sup_{k\in\mathbb{N}} |r_k x_k + s_{k-1}x_{k-1}|^{p_k} < \infty\right\},\\
c(\widetilde{B},p) &:= \left\{x = (x_k) \in \omega : \exists l \in \mathbb{C} \ni \lim_{k\to\infty} |r_k x_k + s_{k-1}x_{k-1} - l|^{p_k} = 0\right\},\\
c_0(\widetilde{B},p) &:= \left\{x = (x_k) \in \omega : \lim_{k\to\infty} |r_k x_k + s_{k-1}x_{k-1}|^{p_k} = 0\right\}.
\end{aligned}$$

It is natural that one can reexpress the sequence spaces $\ell_\infty(\widetilde{B},p)$, $c(\widetilde{B},p)$ and $c_0(\widetilde{B},p)$ with the notation (1.3), as follows:

$$\ell_\infty(\widetilde{B},p) = [\ell_\infty(p)]_{\widetilde{B}}, \quad c(\widetilde{B},p) = [c(p)]_{\widetilde{B}} \quad \text{and} \quad c_0(\widetilde{B},p) = [c_0(p)]_{\widetilde{B}}.$$

Now, we start with the following theorem which is basic for this subsection:

Theorem 3.1.1. *$\ell_\infty(\widetilde{B},p)$, $c(\widetilde{B},p)$ and $c_0(\widetilde{B},p)$ are the complete linear metric spaces paranormed by g_6, defined by*

$$g_6(x) = \sup_{k\in\mathbb{N}} |r_k x_k + s_{k-1}x_{k-1}|^{p_k/M}.$$

Proof. To avoid the repetition of the similar statements, the proof is only given for the space $c_0(\widetilde{B},p)$. Since $c_0(p)$ is a linear space and $B(\widetilde{r},\widetilde{s})$ is a triangle matrix, one can easily show that $c_0(\widetilde{B},p)$ is a linear space with the coordinatewise addition and scalar multiplication.

It is clear that $g_6(\theta) = 0$, $g_6(x) \geq 0$ for all $x \in c_0(\widetilde{B},p)$ and $g_6(-x) = g_6(x)$.

Let (x^n) be any sequence of the points $c_0(\widetilde{B},p)$ such that $g_6(x^n - x) \to 0$, as $n \to \infty$ and (α_n) is a sequence of scalars with $\alpha_n \to \alpha$, as $n \to \infty$. Then, since the inequality

$$g_6(x^n) \leq g_6(x) + g_6(x^n - x)$$

holds by subadditivity of the function g_6, $\{g_6(x^n)\}$ is bounded and we thus have

$$\begin{aligned} g_6(\alpha_n x^n - \alpha x) &= \sup_{k\in\mathbb{N}} \left| r_k(\alpha_n x_k^n - \alpha x_k) + s_{k-1}\left(\alpha_n x_{k-1}^n - \alpha x_{k-1}\right)\right|^{p_k/M} \\ &\leq |(\alpha_n - \alpha)|\, g_6(x^n) + |\alpha|\, g_6(x^n - x) \end{aligned}$$

which tends to zero, as $n \to \infty$. That is to say that the scalar multiplication is continuous. Hence, g_6 is a paranorm on the space $c_0(\widetilde{B}, p)$.

It remains to prove the completeness of the space $c_0(\widetilde{B}, p)$. Let (x^i) be any Cauchy sequence in $c_0(\widetilde{B}, p)$, where $x^i = \{x_0^{(i)}, x_1^{(i)}, x_2^{(i)}, \dots\}$. Then, for a given $\varepsilon > 0$ there exists a positive integer $n_0(\varepsilon)$ such that $g_6(x^i - x^j) < \varepsilon$ for all $i, j \geq n_0(\varepsilon)$. We obtain by using definition of g_6 for each fixed $k \in \mathbb{N}$ that

$$\begin{aligned} \left|\{B(\widetilde{r}, \widetilde{s})x^i\}_k - \{B(\widetilde{r}, \widetilde{s})x^j\}_k\right| &\leq \sup_{k\in\mathbb{N}} \left|\{B(\widetilde{r}, \widetilde{s})x^i\}_k - \{B(\widetilde{r}, \widetilde{s})x^j\}_k\right|^{p_k/M} \\ &< \varepsilon \end{aligned} \tag{3.2}$$

for every $i, j \geq n_0(\varepsilon)$ which leads us to the fact that

$$\{\{B(\widetilde{r}, \widetilde{s})x^0\}_k, \{B(\widetilde{r}, \widetilde{s})x^1\}_k, \{B(\widetilde{r}, \widetilde{s})x^2\}_k, \dots\}$$

is a Cauchy sequence of complex numbers for every fixed $k \in \mathbb{N}$. Since $\mathbb{C}$ is complete, it converges, say $\{B(\widetilde{r}, \widetilde{s})x^i\}_k \to \{B(\widetilde{r}, \widetilde{s})x\}_k$, as $i \to \infty$. Using these infinitely many limits $\{B(\widetilde{r}, \widetilde{s})x\}_0$, $\{B(\widetilde{r}, \widetilde{s})x\}_1$, $\{B(\widetilde{r}, \widetilde{s})x\}_2, \dots$, we define the sequence

$$\{\{B(\widetilde{r}, \widetilde{s})x\}_0, \{B(\widetilde{r}, \widetilde{s})x\}_1, \{B(\widetilde{r}, \widetilde{s})x\}_2, \dots\}.$$

We have from (3.2) with $j \to \infty$ that

$$\sup_{k\in\mathbb{N}} \left|\{B(\widetilde{r}, \widetilde{s})x^i\}_k - \{B(\widetilde{r}, \widetilde{s})x\}_k\right|^{p_k/M} \leq \varepsilon \tag{3.3}$$

for every fixed $k \in \mathbb{N}$. Since $x^i = (x_k^i) \in c_0(\widetilde{B}, p)$, $\left|\{B(\widetilde{r}, \widetilde{s})x^i\}_k\right|^{p_k/M} < \varepsilon$ for all $k \in \mathbb{N}$. Therefore, we obtain by (3.3) that

$$\begin{aligned} |\{B(\widetilde{r}, \widetilde{s})x\}_k|^{p_k/M} &\leq \left|\{B(\widetilde{r}, \widetilde{s})x\}_k - \{B(\widetilde{r}, \widetilde{s})x^i\}_k\right|^{p_k/M} + \left|\{B(\widetilde{r}, \widetilde{s})x^i\}_k\right|^{p_k/M} \\ &< 2\varepsilon. \end{aligned}$$

This shows that the sequence $B(\widetilde{r}, \widetilde{s})x$ belongs to the space $c_0(p)$. Since (x^i) is an arbitrary Cauchy sequence, the space $c_0(\widetilde{B}, p)$ is complete.

As an immediate consequence of the fact that any term with negative subscript is equal to naught and $g_6(x) = 0$ if and only if $r_k x_k + s_{k-1} x_{k-1} = 0$ for all $k \in \mathbb{N}$ which yields that $x_k = 0$ for all $k \in \mathbb{N}$, that is, $g_6(x) = 0$ implies $x = \theta$, as desired. This says that the function g_6 is a total paranorm.

This concludes the proof of Theorem 3.1.1. □

It is trivial that the sequence spaces $c_0(\widetilde{B},p)$, $c(\widetilde{B},p)$ and $\ell_\infty(\widetilde{B},p)$ of non-absolute type are linearly isomorphic to the spaces $c_0(p)$, $c(p)$ and $\ell_\infty(p)$, respectively.

Prior to giving the Schauder basis of the spaces $c_0(\widetilde{B},p)$ and $c(\widetilde{B},p)$, we define the Schauder basis for a paranormed space.

Let (X,g) be a paranormed space. A sequence (b_k) of the elements of X is called a *Schauder basis* for X if and only if for each $x \in X$, there exists a unique sequence (α_k) of scalars such that $g\left(x - \sum_{k=0}^{n} \alpha_k b_k\right) \to 0$, as $n \to \infty$. Using Remark 2.1.4, we have:

Proposition 3.1.2. *Let $\alpha_k = \{B(\widetilde{r},\widetilde{s})x\}_k$ and $0 < p_k \leq M < \infty$ for all $k \in \mathbb{N}$. Define the sequences $a = (a_n)$ and $b^{(k)} = \{b_n^{(k)}(\widetilde{r},\ \widetilde{s})\}_{n\in\mathbb{N}}$ of the elements of the space $c_0(\widetilde{B},p)$ by*

$$a_n = \sum_{k=0}^{n} \frac{(-1)^{n-k}}{r_n} \prod_{j=k}^{n-1} \frac{s_j}{r_j},$$

$$b_n^{(k)}(\widetilde{r},\ \widetilde{s}) := \begin{cases} 0 & , \quad n < k, \\ \frac{(-1)^{n-k}}{r_n} \prod_{j=k}^{n-1} \frac{s_j}{r_j} & , \quad n \geq k \end{cases}$$

for all $k,n \in \mathbb{N}$. Then, the following statements hold:

(i) The sequence $\{b^{(k)}\}$ is a basis for $c_0(\widetilde{B},p)$ and any $x \in c_0(\widetilde{B},p)$ has a unique representation of the form $x = \sum_k \alpha_k b^{(k)}$.

(ii) The set $\{a, b^{(k)}\}$ is a basis for $c(\widetilde{B},p)$ and any $x \in c(\widetilde{B},p)$ has a unique representation of the form

$$x = \xi a + \sum_k (\alpha_k - \xi)\, b^{(k)}, \quad \textit{where} \quad \xi = \lim_{k\to\infty} \{B(\widetilde{r},\widetilde{s})x\}_k.$$

3.1.2 Alpha-, Beta- and Gamma-duals of the Spaces $\ell_\infty(\widetilde{B},p)$, $c(\widetilde{B},p)$ and $c_0(\widetilde{B},p)$

In this subsection, we state and prove the theorem determining the alpha-, beta- and gamma-duals of the spaces $c_0(\widetilde{B},p)$, $c(\widetilde{B},p)$ and $\ell_\infty(\widetilde{B},p)$.

Theorem 3.1.3. *Define the sets $d_1(p)$, $d_2(p)$, $d_3(p)$, $d_4(p)$, $d_5(p)$, $d_6(p)$ and $d_7(p)$ by*

$$d_1(p) := \bigcup_{N>1} \left\{ (a_k) \in \omega : \sup_{K\in\mathcal{F}} \sum_n \left| \sum_{k\in K} \frac{(-1)^{n-k}}{r_n} \prod_{j=k}^{n-1} \frac{s_j}{r_j} a_n \right| N^{-1/p_k} < \infty \right\},$$

$$d_2(p) := \bigcup_{N>1} \left\{ (a_k) \in \omega : \sum_k \left| \sum_{i=k}^{n} \frac{(-1)^i}{r_n} \prod_{j=k}^{i-1} \frac{s_j}{r_j} a_i \right| N^{-1/p_k} < \infty \right\},$$

$$d_3(p) := \bigcup_{N>1}\left\{(a_k)\in\omega : \left(\sum_{i=k}^{n}\frac{(-1)^i}{r_n}\prod_{j=k}^{i-1}\frac{s_j}{r_j}a_iN^{-1/p_k}\right)\in\ell_\infty\right\},$$

$$d_4(p) := \bigcup_{N>1}\left\{(a_k)\in\omega : \sup_{n\in\mathbb{N}}\sum_k\left|\sum_{i=k}^{n}\frac{(-1)^i}{r_n}\prod_{j=k}^{i-1}\frac{s_j}{r_j}a_i\right|N^{-1/p_k}<\infty\right\},$$

$$d_5(p) := \bigcap_{N>1}\left\{(a_k)\in\omega : \sup_{K\in\mathcal{F}}\sum_n\left|\sum_{k\in K}\frac{(-1)^{n-k}}{r_n}\prod_{j=k}^{n-1}\frac{s_j}{r_j}a_n\right|N^{1/p_k}<\infty\right\},$$

$$d_6(p) := \bigcap_{N>1}\left\{(a_k)\in\omega : \sup_{n\in\mathbb{N}}\sum_k\left|\sum_{i=k}^{n}\frac{(-1)^i}{r_n}\prod_{j=k}^{i-1}\frac{s_j}{r_j}a_i\right|N^{1/p_k}<\infty\right\},$$

$$d_7(p) := \left\{(a_k)\in\omega : \sum_n\left|\sum_k\frac{(-1)^{n-k}}{r_n}\prod_{j=k}^{n-1}\frac{s_j}{r_j}a_n\right|<\infty\right\}.$$

Then, we have the following:

$$\begin{aligned}
[c_0(\widetilde{B},p)]^\alpha &= d_1(p), & [c_0(\widetilde{B},p)]^\beta &= d_2(p)\cap d_3(p)\\
[c(\widetilde{B},p)]^\alpha &= d_1(p)\cap d_7(p), & [c(\widetilde{B},p)]^\beta &= d_2(p)\cap d_3(p)\cap cs\\
[\ell_\infty(\widetilde{B},p)]^\alpha &= d_5(p), & [\ell_\infty(\widetilde{B},p)]^\beta &= d_6(p)\cap cs\\
[c_0(\widetilde{B},p)]^\gamma &= d_4(p), & [c(\widetilde{B},p)]^\gamma &= d_4(p)\cap bs\\
[\ell_\infty(\widetilde{B},p)]^\gamma &= d_6(p).
\end{aligned}$$

Proof. We give the proof only for the space $c_0(\widetilde{B},p)$. Let $a=(a_n)\in\omega$ and define the matrix $C=(c_{nk})$ via the sequence $a=(a_n)$ by

$$c_{nk} = \begin{cases}\frac{(-1)^{n-k}}{r_n}\prod_{j=k}^{n-1}\frac{s_j}{r_j}a_n & , \quad 0\le k\le n,\\[2ex] 0 & , \quad k>n\end{cases}$$

for all $n,k\in\mathbb{N}$. Taking into account the relation (3.1) one can derive by straightforward calculation that

$$a_nx_n = a_n\left\{B^{-1}(\widetilde{r},\widetilde{s})y\right\}_n = \sum_{k=0}^{n}\frac{(-1)^{n-k}}{r_n}\prod_{j=k}^{n-1}\frac{s_j}{r_j}y_ka_n = (Cy)_n \qquad (3.4)$$

for all $n\in\mathbb{N}$. We therefore observe by (3.4) that $ax=(a_nx_n)\in\ell_1$ whenever $x=(x_k)\in c_0(\widetilde{B},p)$ if and only if $Cy\in\ell_1$ whenever $y=(y_k)\in c_0(p)$. This means that the sequence $a=(a_n)$ is in α-dual of the spaces $c_0(\widetilde{B},p)$ if and only if $C\in(c_0(p):\ell_1)$. Then we derive by Part (iv) of Lemma 3.1.8 with $q_n=1$ for all $n\in\mathbb{N}$ that $a\in[c_0(\widetilde{B},p)]^\alpha$ if and only if

$$\sup_{K\in\mathcal{F}}\sum_n\left|\sum_{k\in K}c_{nk}\right|N^{-1/p_k}<\infty$$

which leads to the consequence that $[c_0(\widetilde{B},p)]^{\alpha} = d_1(p)$.

Consider the equality

$$\sum_{k=0}^{n} a_k x_k = \sum_{k=0}^{n} \left[\sum_{j=0}^{k} \frac{(-1)^{k-j}}{r_k} \prod_{i=j}^{k-1} \frac{s_i}{r_i} y_j \right] a_k \tag{3.5}$$

$$= \sum_{k=0}^{n} \sum_{i=k}^{n} \frac{(-1)^{i}}{r_i} \prod_{j=k}^{i-1} \frac{s_j}{r_j} a_i y_k$$

$$= (Dy)_n,$$

where $D = (d_{nk})$ is defined by

$$d_{nk} = \begin{cases} \sum\limits_{i=k}^{n} \frac{(-1)^{i}}{r_i} \prod\limits_{j=k}^{i-1} \frac{s_j}{r_j} a_i & , \quad 0 \leq k \leq n, \\ 0 & , \quad k > n \end{cases} \tag{3.6}$$

for all $n, k \in \mathbb{N}$. Thus we deduce from Part (ii) of Lemma 3.1.8 with $q_n = 1$ for all $n \in \mathbb{N}$ and (3.5) that $ax = (a_n x_n) \in cs$ whenever $x = (x_k) \in c_0(\widetilde{B},p)$ if and only if $Dy \in c$ whenever $y = (y_k) \in c_0(p)$. This means that the sequence $a = (a_n)$ is in β-dual of the space $c_0(\widetilde{B},p)$ if and only if $D \in (c_0(p) : c)$, where $D = (d_{nk})$ is defined by (3.6). Therefore, we derive by Part (ii) of Lemma 3.1.8 with $q_n = 1$ for all $n \in \mathbb{N}$ and some $N \in \mathbb{N}_2$ that

$$\sum_{k=0}^{n} \left| \sum_{i=k}^{n} \frac{(-1)^{n-i}}{r_i} \prod_{j=k}^{i-1} \frac{s_j}{r_j} a_i \right| N^{-1/p_k} < \infty$$

and there exists $\alpha_k \in \mathbb{C}$ for all $k \in \mathbb{N}$ such that

$$\sup_{n \in \mathbb{N}} \left| \sum_{k=0}^{n} \sum_{i=k}^{n} \frac{(-1)^{i}}{r_i} \prod_{j=k}^{i-1} \frac{s_j}{r_j} a_i - \alpha_k \right| N^{-1/p_k} < \infty$$

which shows that $[c_0(\widetilde{B},p)]^{\beta} = d_2(p) \cap d_3(p)$.

We deduce from Part (i) of Lemma 3.1.8 that $ax = (a_n x_n) \in bs$ whenever $x = (x_k) \in c_0(\widetilde{B},p)$ if and only if $Dy \in \ell_\infty$ whenever $y = (y_k) \in c_0(p)$. This means that the sequence $a = (a_n)$ is in γ-dual of the space $c_0(\widetilde{B},p)$ if and only if $D \in (c_0(p) : \ell_\infty)$. Therefore, we attain by Part (i) of Lemma 3.1.8 with $q_n = 1$ for all $n \in \mathbb{N}$ that $[c_0(\widetilde{B},p)]^{\gamma} = d_4(p)$. □

3.1.3 Matrix Transformations

Let λ denote any of the spaces c_0, c or ℓ_∞ and μ be any given sequence space. In the present subsection, we characterize the classes $(\lambda(\widetilde{B},p) : \mu)$ and

$(\mu : \lambda(\widetilde{B}, p))$ of infinite matrices. Moreover, we give the characterization of matrix transformations from the space $\lambda(\widetilde{B}, p)$ into the spaces f and f_0. For brevity, the following notations will be used throughout

$$\widetilde{a}_{nk} = \sum_{i=k}^{\infty} \frac{(-1)^i}{r_i} \prod_{j=k}^{i-1} \frac{s_j}{r_j} a_{ni} \text{ and } \widetilde{a}(n,k,m) = \frac{1}{m+1} \sum_{i=0}^{m} \widetilde{a}_{n+i,k}$$

for all $k, m, n \in \mathbb{N}$. We also use these notations with other letters.

Let us suppose that the infinite matrices $E = (e_{nk})$ and $F = (f_{nk})$ transform the sequences $x = (x_k)$ and $y = (y_k)$ which are connected with the relation (3.1) to the sequences $r = (r_n)$ and $s = (s_n)$, respectively, i.e.,

$$r_n = (Ex)_n = \sum_k e_{nk} x_k \text{ for all } n \in \mathbb{N}, \tag{3.7}$$

$$s_n = (Fy)_n = \sum_k f_{nk} y_k \text{ for all } n \in \mathbb{N}. \tag{3.8}$$

One can immediately deduce here that the method F is applied to the $B(\widetilde{r}, \widetilde{s})$-transform of the sequence $x = (x_k)$ while the method E is directly applied to the terms of the sequence $x = (x_k)$. So, the methods E and F are essentially different.

Let us assume that the matrix product $FB(\widetilde{r}, \widetilde{s})$ exists which is a much weaker assumption than the conditions on the matrix F belonging to any matrix class, in general. We shall say in this situation that the methods E and F in (3.7) and (3.8) are the *pair of summability methods*, shortly PSM, if r_n becomes s_n (or s_n becomes r_n) under the application of the formal summation by parts. This leads us to the fact that $FB(\widetilde{r}, \widetilde{s})$ exists and is equal to E and $[FB(\widetilde{r}, \widetilde{s})]x = F[B(\widetilde{r}, \widetilde{s})x]$ formally holds, if one side exists. This statement is equivalent to the relation

$$e_{nk} = r_k f_{nk} + s_k f_{n,k+1} \text{ or equivalently } f_{nk} = \sum_{i=k}^{\infty} \frac{(-1)^i}{r_i} \prod_{j=k}^{i-1} \frac{s_j}{r_j} e_{ni} \tag{3.9}$$

for all $k, n \in \mathbb{N}$. Then, one can see by following the similar way used in Başar's monograph [21, pp. 163-164] for the PSM's that s_n reduces to r_n:

$$\begin{aligned} s_n = \sum_k f_{nk} y_k &= \sum_k f_{nk} (r_k x_k + s_{k-1} x_{k-1}) \\ &= \sum_k (r_k f_{nk} + s_k f_{n,k+1}) x_k = r_n. \end{aligned}$$

But the order of summation may not be reversed which means that the methods E and F are not necessarily equivalent.

We now give the following basic theorem related to the matrix mappings on the sequence spaces $c_0(\widetilde{B}, p)$, $c(\widetilde{B}, p)$ and $\ell_\infty(\widetilde{B}, p)$:

Theorem 3.1.4. *Let the matrices $E = (e_{nk})$ and $F = (f_{nk})$ be connected with the relation (3.9), λ be any of the spaces c_0, c and ℓ_∞, and μ be any given sequence space. Then $E \in (\lambda(\widetilde{B}, p) : \mu)$ if and only if $F \in (\lambda(p) : \mu)$ and*

$$F^n \in (\lambda(p) : c) \tag{3.10}$$

for every fixed $n \in \mathbb{N}$, where $F^n = \left(f_{mk}^{(n)}\right)$ with

$$f_{mk}^{(n)} = \begin{cases} \sum_{i=k}^{m} \frac{(-1)^i}{r_i} \prod_{j=k}^{i-1} \frac{s_j}{r_j} e_{ni} & , \quad 0 \le k \le m, \\ 0 & , \quad k > m, \end{cases}$$

for all $m, k \in \mathbb{N}$.

Proof. Suppose that $E = (e_{nk})$ and $F = (f_{nk})$ are connected with the relation (3.9), and let μ be any given sequence space and keep in mind that the spaces $\lambda(\widetilde{B}, p)$ and $\lambda(p)$ are paranorm isomorphic.

Let $E \in (\lambda(\widetilde{B}, p) : \mu)$ and take any $y \in \lambda(p)$. Then, $FB(\widetilde{r}, \widetilde{s})$ exists and $(e_{nk})_{k\in\mathbb{N}} \in [\lambda(\widetilde{B}, p)]^\beta$ which shows that (3.10) is necessary and $(f_{nk})_{k\in\mathbb{N}} \in [\lambda(p)]^\beta$ for each $n \in \mathbb{N}$. Thus, Fy exists for all $y \in \lambda(p)$ and hence by letting $m \to \infty$ in the equality

$$\sum_{k=0}^{m} e_{nk} x_k = \sum_{k=0}^{m} \left[\sum_{i=k}^{m} \frac{(-1)^i}{r_i} \prod_{j=k}^{i-1} \frac{s_j}{r_j} e_{ni} \right] y_k \tag{3.11}$$

with (3.9) that $Ex = Fy$ which leads us to the fact that $F \in (\lambda(p) : \mu)$.

Conversely, suppose that $F \in (\lambda(p) : \mu)$ and (3.10) holds. Let $y = (y_k) \in \lambda(p)$. Then, we have $(f_{nk})_{k\in\mathbb{N}} \in [\lambda(p)]^\beta$ which gives together with (3.10) that $(e_{nk})_{k\in\mathbb{N}} \in [\lambda(\widetilde{B}, p)]^\beta$ for each $n \in \mathbb{N}$. Hence, Ex exists. As a consequence of the equality (3.11) with $m \to \infty$ that $Fy = Ex$ this proves that $E \in (\lambda(\widetilde{B}, p) : \mu)$. □

By changing the roles of the spaces $\lambda(\widetilde{B}, p)$ and $\lambda(p)$ with μ in Theorem 3.1.4, we have:

Theorem 3.1.5. *Suppose that the elements of the infinite matrices $A = (a_{nk})$ and $C = (c_{nk})$ are connected with the relation*

$$c_{nk} = r_n a_{nk} + s_{n-1} a_{n-1,k} \quad \textit{for all} \quad k, n \in \mathbb{N},$$

λ be any of the spaces c_0, c or ℓ_∞, and μ be any given sequence space. Then, $A = (a_{nk}) \in (\mu : \lambda(\widetilde{B}, p))$ if and only if $C \in (\mu : \lambda(p))$.

Proof. Let $z = (z_k) \in \mu$ and consider the following equality

$$\sum_{k=0}^{m} c_{nk} z_k = \sum_{k=0}^{m} (r_n a_{nk} + s_{n-1} a_{n-1,k})\, z_k \quad \text{for all} \quad m, n \in \mathbb{N}$$

which yields as $m \to \infty$ that $(Cz)_n = \{B(\widetilde{r}, \widetilde{s})Az\}_n$ for all $n \in \mathbb{N}$. Therefore, one can observe from here that $Az \in \lambda(\widetilde{B}, p)$ whenever $z \in \mu$ if and only if $Cz \in \lambda(p)$ whenever $z \in \mu$.

This completes the proof. □

It is easy to see that Theorem 3.1.4 and Theorem 3.1.5 have lots of outcomes depending on the choice of the sequence spaces λ and μ, and the sequence $p = (p_k)$. Whence by Theorem 3.1.4 and Theorem 3.1.5, the necessary and sufficient conditions for $(\lambda(\widetilde{B}, p) : \mu)$ and $(\mu : \lambda(\widetilde{B}, p))$ may be derived by replacing the entries of E and A by those of the entries of $F = EB^{-1}(\widetilde{r}, \widetilde{s})$ and $C = B(\widetilde{r}, \widetilde{s})A$, respectively, where the necessary and sufficient conditions on the matrices F and C are read from the concerning results in the existing literature.

The necessary and sufficient conditions on the matrix mappings from Maddox's spaces to the sequence spaces f and f_0 are given by the following lemma due to Nanda [109]:

Lemma 3.1.6. *Let $A = (a_{nk})$ be an infinite matrix. Then, the following statements hold:*

(i) $A \in (c_0(p) : f)$ if and only if (1.11) and (1.12) hold.

(ii) $A \in (\ell_\infty(p) : f_0)$ if and only if (1.12) holds and

$$\exists B > 1 \ni \lim_{m \to \infty} \sum_k |a(n, k, m)| B^{1/p_k} = 0 \text{ uniformly in } n. \tag{3.12}$$

Let (q_n) be a non-decreasing bounded sequence of positive real numbers, K denotes the finite subsets of $\mathbb{N}$, and also B and M denote the natural numbers. Finally, the sets K_1 and K_2 are defined by $K_1 = \{k \in \mathbb{N} : p_k \leq 1\}$ and $K_2 = \{k \in \mathbb{N} : p_k > 1\}$.

The matrix transformations between Maddox's sequence spaces $\ell(p)$, $c_0(p)$, $c(p)$ and $\ell_\infty(p)$ were characterized by Grosse-Erdmann [64], and some of them are given by the following lemmas:

Lemma 3.1.7. *Let $A = (a_{nk})$ be an infinite matrix. Then, the following statements hold:*

(i) $A \in (\ell_\infty(p) : \ell_\infty(q))$ if and only if

$$\forall B > 1 \ni \sup_{n \in \mathbb{N}} \left(\sum_k |a_{nk}| B^{1/p_k} \right)^{q_n} < \infty. \tag{3.13}$$

(ii) $A \in (\ell_\infty(p) : c(q))$ if and only if

$$\forall B > 1 \ni \sup_{n \in \mathbb{N}} \sum_k |a_{nk}| B^{1/p_k} < \infty, \tag{3.14}$$

$$\exists (\alpha_k) \in \omega, \ \forall B > 1 \ni \lim_{n \to \infty} \left(\sum_k |a_{nk} - \alpha_k| B^{1/p_k} \right)^{q_n} = 0. \tag{3.15}$$

(iii) $A \in (\ell_\infty(p) : c_0(q))$ *if and only if*

$$\forall B > 1 \ni \lim_{n\to\infty} \left(\sum_k |a_{nk}| B^{1/p_k} \right)^{q_n} = 0. \tag{3.16}$$

(iv) $A \in (\ell_\infty(p) : \ell(q))$ *if and only if*

$$\sup_{K\in\mathcal{F}} \sum_n \left| \sum_{k\in K} a_{nk} B^{1/p_k} \right|^{q_n} < \infty \tag{3.17}$$

for all $q_n \geq 1$ *and for all* $B > 1$.

Lemma 3.1.8. *Let* $A = (a_{nk})$ *be an infinite matrix. Then, the following statements hold:*

(i) $A \in (c_0(p) : \ell_\infty(q))$ *if and only if*

$$\exists B > 1 \ni \sup_{n\in\mathbb{N}} \left(\sum_k |a_{nk}| B^{-1/p_k} \right)^{q_n} < \infty. \tag{3.18}$$

(ii) $A \in (c_0(p) : c(q))$ *if and only if*

$$\exists B > 1 \ni \sup_{n\in\mathbb{N}} \sum_k |a_{nk}| B^{-1/p_k} < \infty, \tag{3.19}$$

$$\forall M, \exists B > 1 \ni \sup_{n\in\mathbb{N}} \sum_{k\in K_2} |a_{nk} - \alpha_k| M^{1/q_n} B^{-1/p_k} < \infty, \tag{3.20}$$

$$\exists (\alpha_k) \in \omega \ni \lim_{n\to\infty} |a_{nk} - \alpha_k|^{q_n} = 0. \tag{3.21}$$

(iii) $A \in (c_0(p) : c_0(q))$ *if and only if (3.20) holds with* $\alpha_k = 0$ *for all* $k \in \mathbb{N}$ *and*

$$\lim_{n\to\infty} |a_{nk}|^{q_n} = 0 \quad \text{for each fixed} \quad k \in \mathbb{N}. \tag{3.22}$$

(iv) $A \in (c_0(p) : \ell(q))$ *if and only if*

$$\exists N > 1 \ni \sup_{K\in\mathcal{F}} \sum_n \left| \sum_{k\in K} a_{nk} B^{-1/p_k} \right|^{q_n} < \infty \quad \text{for all} \quad q_n \geq 1. \tag{3.23}$$

Lemma 3.1.9. *Let* $A = (a_{nk})$ *be an infinite matrix. Then, the following statements hold:*

(i) $A \in (c(p) : \ell_\infty(q))$ *if and only if (3.18) holds and*

$$\sup_{n\in\mathbb{N}} \left| \sum_k a_{nk} \right|^{q_n} < \infty. \tag{3.24}$$

(ii) $A \in (c(p) : c(q))$ *if and only if (3.19), (3.20) and (3.21) hold and*

$$\exists \alpha \in \mathbb{C} \ni \lim_{n\to\infty} \left|\sum_k a_{nk} - \alpha\right|^{q_n} = 0. \tag{3.25}$$

(iii) $A \in (c(p) : c_0(q))$ *if and only if (3.22), holds and*

$$\forall M, \exists B > 1 \ni \sup_{n\in\mathbb{N}} \sum_{k\in K_2} |a_{nk}|\, M^{1/q_n} B^{-1/p_k} < \infty, \tag{3.26}$$

$$\lim_{n\to\infty} \left|\sum_k a_{nk}\right|^{q_n} = 0. \tag{3.27}$$

(iv) $A \in (c(p) : \ell(q))$ *if and only if (3.23) holds and*

$$\sum_n \left|\sum_k a_{nk}\right|^{q_n} < \infty \ \ \textit{for all} \ \ q_n \geq 1. \tag{3.28}$$

Lemma 3.1.10. *Let* $A = (a_{nk})$ *be an infinite matrix. Then, the following statements hold:*

(i) $A \in (\ell(p) : c(q))$ *if and only if (3.21) holds and*

$$\sup_{n\in\mathbb{N}} \sup_{k\in K_1} |a_{nk}|^{p_k} < \infty, \tag{3.29}$$

$$\exists B > 1 \ni \sup_{n\in\mathbb{N}} \sum_{k\in K_2} |a_{nk} B^{-1}|^{p'_k} < \infty, \tag{3.30}$$

$$\exists\, (\alpha_k) \in \omega \text{ and } \forall M > 1 \ni \sup_{n\in\mathbb{N}} \sup_{k\in K_1} \left(|a_{nk} - \alpha_k|\, M^{1/q_n}\right)^{p_k} < \infty, \tag{3.31}$$

$$\exists\, (\alpha_k) \in \omega,\ \forall M, \exists B > 1 \ni \sup_{n\in\mathbb{N}} \sum_{k\in K_2} \left(|a_{nk} - \alpha_k|\, M^{1/q_n} B^{-1}\right)^{p'_k} < \infty. \tag{3.32}$$

(ii) $A \in (\ell(p) : c_0(q))$ *if and only if (3.22) holds and for all* $M > 1$,

$$\sup_{n\in\mathbb{N}} \sup_{k\in K_1} \left|a_{nk} M^{1/q_n}\right|^{p_k} < \infty \ \ \textit{for all} \ \ M > 1, \tag{3.33}$$

$$\exists B > 1 \ni \sup_{n\in\mathbb{N}} \sum_{k\in K_2} \left|a_{nk} M^{1/q_n} B^{-1}\right|^{p'_k} < \infty. \tag{3.34}$$

(iii) $A \in (\ell(p) : \ell_\infty(q))$ *if and only if*

$$\exists M > 1 \ni \sup_{n\in\mathbb{N}} \sup_{k\in K_2} \left|a_{nk} M^{-1/q_n}\right|^{p_k} < \infty, \tag{3.35}$$

$$\exists M > 1 \ni \sup_{n\in\mathbb{N}} \sum_{k\in K_2} \left|a_{nk} M^{-1/q_n}\right|^{p'_k} < \infty. \tag{3.36}$$

Now, we can quote our theorems without proof on the characterization of some matrix classes concern with the sequence spaces f, f_0, $c_0(\widetilde{B},p)$, $c(\widetilde{B},p)$ and $\ell_\infty(\widetilde{B},p)$.

Theorem 3.1.11. *Let $A=(a_{nk})$ be an infinite matrix. Then, the following statements hold:*

(i) $A \in (c(\widetilde{B},p):f)$ if and only if (1.11)-(1.13) hold with $\widetilde{a}(n,k,m)$ instead of $a(n,k,m)$ and (3.10) also holds with $\lambda = c$.

(ii) $A \in (c_0(\widetilde{B},p):f)$ if and only if (1.11) and (1.12) hold with $\widetilde{a}(n,k,m)$ instead of $a(n,k,m)$ and (3.10) also holds with $\lambda = c_0$.

(iii) $A \in (\ell_\infty(\widetilde{B},p):f)$ if and only if (1.11), (1.12) and (1.15) hold with $\widetilde{a}(n,k,m)$ instead of $a(n,k,m)$ and (3.10) also holds with $\lambda = \ell_\infty$.

(iv) $A \in (\ell_\infty(\widetilde{B},p):f_0)$ if and only if (1.12) and (3.12) hold with $\widetilde{a}(n,k,m)$ instead of $a(n,k,m)$ and (3.10) also holds with $\lambda = \ell_\infty$.

Theorem 3.1.12. *Let $A=(a_{nk})$ be an infinite matrix. Then, the following statements hold:*

(i) $A \in (c_0(\widetilde{B},p):\ell(q))$ if and only if (3.23) also holds with $\widetilde{a}_{nk}$ instead of a_{nk} and (3.10) also holds with $\lambda = c_0$.

(ii) $A \in (c_0(\widetilde{B},p):c(q))$ if and only if (3.19), (3.20) and (3.21) hold with $\widetilde{a}_{nk}$ instead of a_{nk} and (3.10) also holds with $\lambda = c_0$.

(iii) $A \in (c_0(\widetilde{B},p):\ell_\infty(q))$ if and only if (3.18) also holds with $\widetilde{a}_{nk}$ instead of a_{nk} and (3.10) also holds with $\lambda = c_0$.

Theorem 3.1.13. *Let $A=(a_{nk})$ be an infinite matrix. Then, the following statements hold:*

(i) $A \in (c(\widetilde{B},p):\ell(q))$ if and only if (3.23) and (3.28) hold with $\widetilde{a}_{nk}$ instead of a_{nk} and (3.10) also holds with $\lambda = c$.

(ii) $A \in (c(\widetilde{B},p):c(q))$ if and only if (3.19), (3.20), (3.21) and (3.25) hold with $\widetilde{a}_{nk}$ instead of a_{nk} and (3.10) also holds with $\lambda = c$.

(iii) $A \in (c(\widetilde{B},p):\ell_\infty(q))$ if and only if (3.18) and (3.24) hold with $\widetilde{a}_{nk}$ instead of a_{nk} and (3.10) also holds with $\lambda = c$.

Theorem 3.1.14. *Let $A=(a_{nk})$ be an infinite matrix. Then, the following statements hold:*

(i) $A \in (\ell_\infty(\widetilde{B},p):\ell(q))$ if and only if (3.17) also holds with $\widetilde{a}_{nk}$ instead of a_{nk} and (3.10) also holds with $\lambda = \ell_\infty$.

(ii) $A \in (\ell_\infty(\widetilde{B},p) : c_0(q))$ *if and only if (3.16) also holds with* $\widetilde{a}_{nk}$ *instead of* a_{nk} *and (3.10) also holds with* $\lambda = \ell_\infty$.

(iii) $A \in (\ell_\infty(\widetilde{B},p) : c(q))$ *if and only if (3.14) and (3.15) hold with* $\widetilde{a}_{nk}$ *instead of* a_{nk} *and (3.10) also holds with* $\lambda = \ell_\infty$.

(iv) $A \in (\ell_\infty(\widetilde{B},p) : \ell_\infty(q))$ *if and only if (3.13) also holds with* $\widetilde{a}_{nk}$ *instead of* a_{nk} *and (3.10) also holds with* $\lambda = \ell_\infty$.

Theorem 3.1.15. *Let* $A = (a_{nk})$ *be an infinite matrix. Then, the following statements hold:*

(i) $A \in (c_0(p) : c(\widetilde{B},q))$ *if and only if (3.19), (3.20) and (3.21) hold with* c_{nk} *instead of* a_{nk}.

(ii) $A \in (c_0(p) : \ell_\infty(\widetilde{B},q))$ *if and only if (3.18) also holds with* c_{nk} *instead of* a_{nk}.

Theorem 3.1.16. *Let* $A = (a_{nk})$ *be an infinite matrix. Then, the following statements hold:*

(i) $A \in (c(p) : c(\widetilde{B},q))$ *if and only if (3.19), (3.20), (3.21) and (3.25) hold with* c_{nk} *instead of* a_{nk}.

(ii) $A \in (c(p) : \ell_\infty(\widetilde{B},q))$ *if and only if (3.18) and (3.24) hold with* c_{nk} *instead of* a_{nk}.

Theorem 3.1.17. *Let* $A = (a_{nk})$ *be an infinite matrix. Then, the following statements hold:*

(i) $A \in (\ell_\infty(p) : c_0(\widetilde{B},q))$ *if and only if (3.16) also holds with* c_{nk} *instead of* a_{nk}.

(ii) $A \in (\ell_\infty(p) : c(\widetilde{B},q))$ *if and only if (3.14) and (3.15) hold with* c_{nk} *instead of* a_{nk}.

(iii) $A \in (\ell_\infty(p) : \ell_\infty(\widetilde{B},q))$ *if and only if (3.13) also holds with* c_{nk} *instead of* a_{nk}.

Theorem 3.1.18. *Let* $A = (a_{nk})$ *be an infinite matrix. Then, the following statements hold:*

(i) $A \in (\ell(p) : c_0(\widetilde{B},q))$ *if and only if (3.22) and (3.33)-(3.34) hold with* c_{nk} *instead of* a_{nk}.

(ii) $A \in (\ell(p) : c(\widetilde{B},q))$ *if and only if (3.21) and (3.29)-(3.32) hold with* c_{nk} *instead of* a_{nk}.

(iii) $A \in (\ell(p) : \ell_\infty(\widetilde{B},q))$ *if and only if (3.35) and (3.36) hold with* c_{nk} *instead of* a_{nk}.

3.2 The Sequence Space $\ell(\widetilde{B},p)$ of Non-absolute Type

In this section, as a continuation of Kirişçi and Başar [73], following Nergiz and Başar [112, 113, 114], we define the linear sequence space $\ell(\widetilde{B},p)$ and prove that it is a complete paranormed space with a Schauder basis. We devote Subsection 3.2.1 to the determination of alpha-, beta- and gamma-duals of the space $\ell(\widetilde{B},p)$. In Subsection 3.2.2, we characterize the classes $(\ell(\widetilde{B},p):\ell_\infty)$, $(\ell(\widetilde{B},p):f)$, $(\ell(\widetilde{B},p):c)$ and $(\ell(\widetilde{B},p):c_0)$ of infinite matrices. Additionally, we obtain the characterizations of some other classes of matrix transformations from the space $\ell(\widetilde{B},p)$ to the Euler, Riesz, difference, etc., sequence spaces by means of Lemma 2.2.30. In the final subsection, we note open problems and further suggestions.

Choudhary and Mishra [40] defined the sequence space $\overline{\ell(p)}$ which consists of all sequences whose S-transforms are in the space $\ell(p)$. Başar and Altay [24] have recently examined the space $bs(p)$ which was formerly defined by Başar in [20] as the set of all series whose sequences of partial sums are in $\ell_\infty(p)$. More recently, Aydın and Başar [17] have studied the space $a^r(u,p)$ which is the domain of the matrix A^r in the sequence space $\ell(p)$, where $0<r<1$. Altay and Başar [4] have studied the sequence space $r^t(p)$ which is derived from the sequence space $\ell(p)$ of Maddox by the Riesz mean R^t. With the notation of (1.3), the spaces $\overline{\ell(p)}, bs(p), a^r(u,p)$ and $r^t(p)$ can be redefined by

$$\overline{\ell(p)}=[\ell(p)]_S,\quad bs(p)=[\ell_\infty(p)]_S,\quad a^r(u,p)=[\ell(p)]_{A^r},\quad r^t(p)=[\ell(p)]_{R^t}.$$

Following Choudhary and Mishra [40], Başar and Altay [24], Altay and Başar [4, 7, 8, 9], Aydın and Başar [17, 14], we introduce the sequence space $\ell(\widetilde{B},p)$ as the set of all sequences whose $B(\widetilde{r},\widetilde{s})$-transforms are in the space $\ell(p)$, that is

$$\ell(\widetilde{B},p):=\left\{(x_k)\in\omega:\sum_k|s_{k-1}x_{k-1}+r_kx_k|^{p_k}<\infty\right\},\quad (0<p_k\le H<\infty).$$

It is trivial that in the case $p_k=p$ for all $k\in\mathbb{N}$, the sequence space $\ell(\widetilde{B},p)$ is reduced to the sequence space $\widetilde{\ell_p}$ which is introduced by Kirişçi and Başar [73]. With the notation of (1.3), we can redefine the space $\ell(\widetilde{B},p)$ as $\ell(\widetilde{B},p):=[\ell(p)]_{B(\widetilde{r},\widetilde{s})}$. Since the spaces $\ell(p)$ and $\ell(\widetilde{B},p)$ are linearly paranorm isomorphic by Corollary 3.2.4, one can easily observe that $x=(x_k)\in\ell(\widetilde{B},p)$ if and only if $y=(y_k)\in\ell(p)$, where the sequences $x=(x_k)$ and $y=(y_k)$ are connected with the relation (3.1).

Now, we begin with the following theorem:

Theorem 3.2.1. *$\ell(\widetilde{B},p)$ is a complete linear metric space paranormed by the paranorm*

$$h(x) = \left(\sum_k |s_{k-1}x_{k-1} + r_k x_k|^{p_k}\right)^{1/M}. \tag{3.1}$$

Proof. It is easy to see that the space $\ell(\widetilde{B},p)$ is linear with respect to the coordinatewise addition and scalar multiplication. Therefore, we first show that it is a paranormed space with the paranorm h defined by (3.1).

It is clear that $h(\theta) = 0$, where $\theta = (0,0,0,\ldots)$ and $h(x) = h(-x)$ for all $x \in \ell(\widetilde{B},p)$.

Let $x, y \in \ell(\widetilde{B},p)$. Then, we derive by Minkowski's inequality that

$$\begin{aligned} h(x+y) &= \left[\sum_k |s_{k-1}(x_{k-1}+y_{k-1}) + r_k(x_k+y_k)|^{p_k}\right]^{1/M} \\ &= \left\{\sum_k \left[|s_{k-1}(x_{k-1}+y_{k-1}) + r_k(x_k+y_k)|^{p_k/M}\right]^M\right\}^{1/M} \\ &\leq \left(\sum_k |s_{k-1}x_{k-1} + r_k x_k|^{p_k}\right)^{1/M} + \left(\sum_k |s_{k-1}y_{k-1} + r_k y_k|^{p_k}\right)^{1/M} \\ &= h(x) + h(y). \end{aligned}$$

Let (λ_n) be a sequence of scalars with $\lambda_n \to \lambda$, as $n \to \infty$ and $\left(x^{(n)}\right)_{n=0}^{\infty}$ be a sequence of elements $x^{(n)} \in \ell(\widetilde{B},p)$ with $h\left(x^{(n)} - x\right) \to 0$, as $n \to \infty$. Therefore, we observe that

$$\begin{aligned} & h\left(\lambda_n x^{(n)} - \lambda x\right) \leq \\ & h\left[(\lambda_n - \lambda)\left(x^{(n)} - x\right)\right] + h\left[\lambda\left(x^{(n)} - x\right)\right] + h\left[(\lambda_n - \lambda)x\right]. \end{aligned} \tag{3.2}$$

It follows from $\lambda_n \to \lambda$, as $n \to \infty$ that $|\lambda_n - \lambda| < 1$ for all sufficiently large n. Hence,

$$\lim_{n\to\infty} h\left[(\lambda_n - \lambda)\left(x^{(n)} - x\right)\right] \leq \lim_{n\to\infty} h\left(x^{(n)} - x\right) = 0. \tag{3.3}$$

Furthermore, we have

$$\lim_{n\to\infty} h\left[\lambda\left(x^{(n)} - x\right)\right] \leq \max\{1, |\lambda|^M\} \lim_{n\to\infty} h\left(x^{(n)} - x\right) = 0. \tag{3.4}$$

Also, we have

$$\lim_{n\to\infty} h\left[(\lambda_n - \lambda)x\right] \leq \lim_{n\to\infty} |\lambda_n - \lambda| h(x) = 0. \tag{3.5}$$

Then, we obtain from (3.2), (3.3), (3.4) and (3.5) that $h(\lambda_n x^{(n)} - \lambda x) \to 0$, as $n \to \infty$. This shows that h is a paranorm on $\ell(\widetilde{B}, p)$.

Furthermore, if $h(x) = 0$, then $(\sum_k |s_{k-1}x_{k-1} + r_k x_k|^{p_k})^{1/M} = 0$. Therefore $|s_{k-1}x_{k-1} + r_k x_k|^{p_k} = 0$ for each $k \in \mathbb{N}$. If we put $k = 0$, since $s_{-1} = 0$ and $r_0 \neq 0$, we have $x_0 = 0$. For $k = 1$, since $x_0 = 0$ we have $x_1 = 0$. Continuing in this way, we obtain $x_k = 0$ for all $k \in \mathbb{N}$. That is, $x = \theta$. This shows that h is a total paranorm.

Now, we show that $\ell(\widetilde{B}, p)$ is complete. Let (x^n) be any Cauchy sequence in $\ell(\widetilde{B}, p)$, where $x^n = \{x_0^{(n)}, x_1^{(n)}, x_2^{(n)}, \ldots\}$. Here and after, for short we write $\widetilde{B}$ instead of $B(\widetilde{r}, \widetilde{s})$. Then for a given $\varepsilon > 0$, there exists a positive integer $n_0(\varepsilon)$ such that $h(x^n - x^m) < \varepsilon$ for all $n, m > n_0(\varepsilon)$. Since for each fixed $k \in \mathbb{N}$

$$\begin{aligned} \left|\left(\widetilde{B}x^n\right)_k - \left(\widetilde{B}x^m\right)_k\right| &\leq \left[\sum_k \left|\left(\widetilde{B}x^n\right)_k - \left(\widetilde{B}x^m\right)_k\right|^{p_k}\right]^{1/M} \\ &= h\left(x^n - x^m\right) < \varepsilon \end{aligned}$$

for every $n, m > n_0(\varepsilon)$, $\{(\widetilde{B}x^0)_k, (\widetilde{B}x^1)_k, (\widetilde{B}x^2)_k, \ldots\}$ is a Cauchy sequence of real numbers for every fixed $k \in \mathbb{N}$. Since $\mathbb{R}$ is complete, it converges, say $(\widetilde{B}x^n)_k \to (\widetilde{B}x)_k$, as $n \to \infty$. Using these infinitely many limits $(\widetilde{B}x)_0, (\widetilde{B}x)_1, (\widetilde{B}x)_2, \ldots$, we define the sequence $\{(\widetilde{B}x)_0, (\widetilde{B}x)_1, (\widetilde{B}x)_2, \ldots\}$. For each $K \in \mathbb{N}$ and $n, m > n_0(\varepsilon)$

$$\left[\sum_{k=0}^{K} \left|\left(\widetilde{B}x^n\right)_k - \left(\widetilde{B}x^m\right)_k\right|^{p_k}\right]^{1/M} \leq h\left(x^n - x^m\right) < \varepsilon.$$

By letting $m, K \to \infty$, one can see for $n > n_0(\varepsilon)$ that

$$h\left(x^n - x\right) = \left[\sum_k \left|\left(\widetilde{B}x^n\right)_k - \left(\widetilde{B}x\right)_k\right|^{p_k}\right]^{1/M} < \varepsilon.$$

This shows that $x^n - x \in \ell(\widetilde{B}, p)$. Since $\ell(\widetilde{B}, p)$ is a linear space, we conclude that $x \in \ell(\widetilde{B}, p)$. It follows that $x^n \to x$, as $n \to \infty$ in $\ell(\widetilde{B}, p)$ thus we have shown that $\ell(\widetilde{B}, p)$ is complete. □

Therefore, one can easily check that the absolute property does not hold on the space $\ell(\widetilde{B}, p)$, that is $h(x) \neq h(|x|)$, where $|x| = (|x_k|)$. This says that $\ell(\widetilde{B}, p)$ is the sequence space of non-absolute type.

Theorem 3.2.2. *Convergence in $\ell(\widetilde{B}, p)$ is stronger than coordinatewise convergence.*

Proof. First we show that $h(x^n - x) \to 0$, as $n \to \infty$ implies $x_k^n \to x_k$, as $n \to \infty$ for every $k \in \mathbb{N}$. We fix k; then we have

$$\lim_{n\to\infty} \left|s_{k-1}x_{k-1}^{(n)} + r_k x_k^{(n)} - s_{k-1}x_{k-1} - r_k x_k\right|^{p_k}$$

$$\leq \lim_{n\to\infty} \sum_k \left| s_{k-1}x_{k-1}^{(n)} + r_k x_k^{(n)} - s_{k-1}x_{k-1} - r_k x_k \right|^{p_k}$$
$$= \lim_{n\to\infty} \left[h\left(x^n - x\right)\right]^M = 0.$$

Hence, we have for $k = 0$ that

$$\lim_{n\to\infty} \left| s_{-1}x_{-1}^{(n)} + r_0 x_0^{(n)} - s_{-1}x_{-1} - r_0 x_0 \right| = 0$$

which gives the fact that $\left|x_0^{(n)} - x_0\right| \to 0$, as $n \to \infty$. Similarly, for each $k \in \mathbb{N}$, we have $\left|x_k^{(n)} - x_k\right| \to 0$, as $n \to \infty$. □

Theorem 3.2.3. *$(\ell_p)_{\widetilde{B}}$ is the linear space under the coordinatewise addition and scalar multiplication which is BK-space with the norm*

$$\|x\| = \left(\sum_k |s_{k-1}x_{k-1} + r_k x_k|^p \right)^{1/p}, \quad \textit{where} \quad 1 \leq p < \infty.$$

Proof. Because the first part of the theorem is a routine verification, we omit details. Since ℓ_p is the BK-space with respect to its usual norm (see [88, pp. 217-218]) and $B(\widetilde{r}, \widetilde{s})$ is a normal matrix, Theorem 4.3.2 of Wilansky [135, p. 61] gives the fact that $(\ell_p)_{\widetilde{B}}$ is the BK-space, where $1 \leq p < \infty$. □

Let us suppose that $1 < p_k \leq s_k$ for all $k \in \mathbb{N}$. Then, it is known that $\ell(p) \subset \ell(s)$ which leads us to the immediate consequence that $\ell(\widetilde{B}, p) \subset \ell(\widetilde{B}, s)$.

With the notation of (3.1), define the transformation T from $\ell(\widetilde{B}, p)$ to $\ell(p)$ by $x \mapsto y = Tx = \widetilde{B}x$. Since T is linear and bijection, we have

Corollary 3.2.4. *The sequence space $\ell(\widetilde{B}, p)$ of non-absolute type is linearly paranorm isomorphic to the space $\ell(p)$, where $0 < p_k \leq H < \infty$ for all $k \in \mathbb{N}$.*

Theorem 3.2.5. *The space $\ell(\widetilde{B}, p)$ has AK.*

Proof. For each $x = (x_k) \in \ell(\widetilde{B}, p)$, we put

$$x^{\langle m \rangle} = \sum_{k=0}^{m} x_k e^{(k)} \quad \text{for all} \quad m \in \mathbb{N}_1.$$

Let $\varepsilon > 0$ and $x \in \ell(\widetilde{B}, p)$ be given. Then, there is $N = N(\varepsilon) \in \mathbb{N}$ such that

$$\sum_{k=N}^{\infty} |s_{k-1}x_{k-1} + r_k x_k|^{p_k} < \varepsilon^M.$$

Then, we have

$$\begin{aligned} h\left(x-x^{\langle m\rangle}\right) &= h\left(x-\sum_{k=1}^{m} x_k e^{(k)}\right) \\ &= \left(\sum_{k=m+1}^{\infty} |s_{k-1}x_{k-1}+r_k x_k|^{p_k}\right)^{1/M} \\ &\leq \left(\sum_{k=N}^{\infty} |s_{k-1}x_{k-1}+r_k x_k|^{p_k}\right)^{1/M} < \varepsilon \end{aligned}$$

for all $m \geq N$. This shows that $x = \sum_k x_k e^{(k)}$.

Now, we have to show that this representation is unique. We assume that $x = \sum_k \lambda_k e^{(k)}$. Then,

$$\begin{aligned} & (|s_{k-1}\lambda_{k-1}+r_k\lambda_k - s_{k-1}x_{k-1} - r_k x_k|^{p_k})^{1/M} \\ & \leq \left(\sum_k |s_{k-1}\lambda_{k-1}+r_k\lambda_k - s_{k-1}x_{k-1} - r_k x_k|^{p_k}\right)^{1/M} \\ & = h(x-x) = 0 \end{aligned}$$

for each $k \in \mathbb{N}$. Hence, $s_{k-1}\lambda_{k-1}+r_k\lambda_k = s_{k-1}x_{k-1}+r_k x_k$ for all $k \in \mathbb{N}$.

For $k=0$, $r_0\lambda_0 = r_0 x_0$. Since $r_0 \neq 0$, we have $\lambda_0 = x_0$.

For $k=1$, $s_0\lambda_0 + r_1\lambda_1 = s_0 x_0 + r_1 x_1$. Since $r_1 \neq 0$, we also have $\lambda_1 = x_1$.

Continuing in this way, we obtain $\lambda_k = x_k$ for all $k \in \mathbb{N}$. Therefore, the representation is unique. □

By Remark 2.1.4, we have:

Corollary 3.2.6. *Let* $0 < p_k \leq H < \infty$ *and* $\alpha_k = (\widetilde{B}x)_k$ *for all* $k \in \mathbb{N}$. *Define the sequence* $b^{(k)} = \{b_n^{(k)}\}_{n\in\mathbb{N}}$ *of the elements of the space* $\ell(\widetilde{B},p)$ *by*

$$b_n^{(k)} = \begin{cases} \frac{(-1)^{n-k}}{r_n}\prod_{j=k}^{n-1}\frac{s_j}{r_j} & , \quad 0 \leq k \leq n, \\ 0 & , \quad k > n \end{cases} \tag{3.6}$$

for every fixed $k \in \mathbb{N}$. *Then, the sequence* $\{b^{(k)}\}_{k\in\mathbb{N}}$ *given by (3.6) is a basis for the space* $\ell(\widetilde{B},p)$ *and any* $x \in \ell(\widetilde{B},p)$ *has a unique representation of the form* $x = \sum_k \alpha_k b^{(k)}$.

3.2.1 The Alpha-, Beta- and Gamma-duals of the Space $\ell(\widetilde{B},p)$

In this subsection, we state and prove the theorems determining the alpha-, beta- and gamma-duals of the sequence space $\ell(\widetilde{B},p)$ of non-absolute type.

We begin with quoting three lemmas which are needed in proving Theorems 3.2.10-3.2.12.

Lemma 3.2.7. [77, (i) and (ii) of Theorem 1] *Let $A = (a_{nk})$ be an infinite matrix. Then, the following statements hold:*

(i) Let $0 < p_k \leq 1$ for all $k \in \mathbb{N}$. Then, $A \in (\ell(p) : \ell_\infty)$ if and only if

$$\sup_{n,k \in \mathbb{N}} |a_{nk}|^{p_k} < \infty. \tag{3.7}$$

(ii) Let $1 < p_k \leq H < \infty$ for all $k \in \mathbb{N}$. Then, $A \in (\ell(p) : \ell_\infty)$ if and only if there exists an integer $M > 1$ such that

$$\sup_{n \in \mathbb{N}} \sum_k \left|a_{nk} M^{-1}\right|^{p'_k} < \infty. \tag{3.8}$$

Lemma 3.2.8. [77, Corollary for Theorem 1] *Let $0 < p_k \leq H < \infty$ for all $k \in \mathbb{N}$. Then, $A = (a_{nk}) \in (\ell(p) : c)$ if and only if (3.7), (3.8) hold, and*

$$\lim_{n \to \infty} a_{nk} = \beta_k \text{ for all } k \in \mathbb{N}. \tag{3.9}$$

Lemma 3.2.9. [64, Theorem 5.1.0] *Let $A = (a_{nk})$ be an infinite matrix. Then, the following statements hold:*

(i) Let $0 < p_k \leq 1$ for all $k \in \mathbb{N}$. Then, $A \in (\ell(p) : \ell_1)$ if and only if

$$\sup_{N \in \mathcal{F}} \sup_{k \in \mathbb{N}} \left| \sum_{n \in N} a_{nk} \right|^{p_k} < \infty.$$

(ii) Let $1 < p_k \leq H < \infty$ for all $k \in \mathbb{N}$. Then, $A \in (\ell(p) : \ell_1)$ if and only if there exists an integer $M > 1$ such that

$$\sup_{N \in \mathcal{F}} \sum_k \left| \sum_{n \in N} a_{nk} M^{-1} \right|^{p'_k} < \infty. \tag{3.10}$$

Because of the case $0 < p_k \leq 1$ may be established in the similar way to the proof of the case $1 < p_k \leq H < \infty$, we omit details of that case and give the proof only for the case $1 < p_k \leq H < \infty$ in Theorems 3.2.10-3.2.12, below.

Theorem 3.2.10. *Define the sets $S_1(p)$ and $S_2(p)$ by*

$$S_1(p) := \bigcup_{B>1} \left\{ (a_n) \in \omega : \sup_{N \in F} \sum_k \left| \sum_{n \in N} \frac{(-1)^{n-k}}{r_n} \prod_{j=k}^{n-1} \frac{s_j}{r_j} a_n M^{-1} \right|^{p'_k} < \infty \right\},$$

$$S_2(p) := \left\{ (a_n) \in \omega : \sup_{N \in F} \sup_{k \in \mathbb{N}} \left| \sum_{n \in N} \frac{(-1)^{n-k}}{r_n} \prod_{j=k}^{n-1} \frac{s_j}{r_j} a_n \right|^{p_k} < \infty \right\}.$$

Then, $\{\ell(\widetilde{B}, p)\}^{\alpha} = \begin{cases} S_1(p) & , \quad 1 < p_k \leq H < \infty \quad \textit{for all} \quad k \in \mathbb{N}, \\ S_2(p) & , \quad 0 < p_k \leq 1 \quad \textit{for all} \quad k \in \mathbb{N}. \end{cases}$

Proof. Let us take any $a = (a_n) \in \omega$. By using (3.1), we obtain that

$$x_n = \sum_{k=0}^{n} \frac{(-1)^{n-k}}{r_n} \prod_{j=k}^{n-1} \frac{s_j}{r_j} y_k$$

for all $n \in \mathbb{N}$ which leads us to

$$a_n x_n = \sum_{k=0}^{n} \frac{(-1)^{n-k}}{r_n} \prod_{j=k}^{n-1} \frac{s_j}{r_j} a_n y_k = (Cy)_n \tag{3.11}$$

for each $n \in \mathbb{N}$, where $C = (c_{nk})$ is defined by

$$c_{nk} = \begin{cases} \frac{(-1)^{n-k}}{r_n} \prod_{j=k}^{n-1} \frac{s_j}{r_j} a_n & , \quad 0 \leq k \leq n \\ 0 & , \quad k > n \end{cases}$$

for all $k, n \in \mathbb{N}$. Thus, we observe by combining (3.11) with the condition (3.10) of Part (ii) of Lemma 3.2.9 that $ax = (a_n x_n) \in \ell_1$ whenever $x = (x_k) \in \ell(\widetilde{B}, p)$ if and only if $Cy \in \ell_1$ whenever $y = (y_k) \in \ell(p)$. This means that $\{\ell(\widetilde{B}, p)\}^{\alpha} = S_1(p)$. □

Theorem 3.2.11. *Define the sets* $S_3(p)$, $S_4(p)$ *and* $S_5(p)$ *by*

$$S_3(p) := \bigcup_{B>1} \left\{ (a_i) \in \omega : \sup_{n} \sum_{k} \left| \sum_{i=k}^{n} \frac{(-1)^{i-k}}{r_i} \prod_{j=k}^{i-1} \frac{s_j}{r_j} a_i M^{-1} \right|^{p'_k} < \infty \right\},$$

$$S_4(p) := \left\{ (a_i) \in \omega : \sum_{i=k}^{\infty} \frac{(-1)^{i-k}}{r_i} \prod_{j=k}^{i-1} \frac{s_j}{r_j} a_i < \infty \right\},$$

$$S_5(p) := \left\{ (a_i) \in \omega : \sup_{n,k \in \mathbb{N}} \left| \sum_{i=k}^{n} \frac{(-1)^{i-k}}{r_i} \prod_{j=k}^{i-1} \frac{s_j}{r_j} a_i \right|^{p_k} < \infty \right\}.$$

Then, $\{\ell(\widetilde{B}, p)\}^{\beta} = \begin{cases} S_3(p) \cap S_4(p) & , \quad 1 < p_k \leq H < \infty \quad \textit{for all} \quad k \in \mathbb{N}, \\ S_4(p) \cap S_5(p) & , \quad 0 < p_k \leq 1 \quad \textit{for all} \quad k \in \mathbb{N}. \end{cases}$

Proof. Define $D = (d_{nk})$ by

$$d_{nk} := \begin{cases} \sum_{i=k}^{n} \frac{(-1)^{i-k}}{r_i} \prod_{j=k}^{i-1} \frac{s_j}{r_j} a_i & , \quad 0 \le k \le n \\ 0 & , \quad k > n \end{cases} \tag{3.12}$$

for all $k, n \in \mathbb{N}$. Take any $a = (a_i) \in \omega$ and consider the equality obtained with (3.1) that

$$\begin{aligned} \sum_{i=0}^{n} a_i x_i &= \sum_{i=0}^{n} \left[\sum_{k=0}^{i} \frac{(-1)^{i-k}}{r_i} \prod_{j=k}^{i-1} \frac{s_j}{r_j} y_k \right] a_i \\ &= \sum_{k=0}^{n} \left[\sum_{i=k}^{n} \frac{(-1)^{i-k}}{r_i} \prod_{j=k}^{i-1} \frac{s_j}{r_j} a_i \right] y_k \\ &= (Dy)_n. \end{aligned} \tag{3.13}$$

Thus, we deduce from Lemma 3.2.8 with (3.13) that $ax = (a_i x_i) \in cs$ whenever $x = (x_i) \in \ell(\widetilde{B}, p)$ if and only if $Dy \in c$ whenever $y = (y_k) \in \ell(p)$. Therefore, we derive from (3.8) and (3.9) that

$$\sup_{n \in \mathbb{N}} \sum_k \left| \sum_{i=k}^{n} \frac{(-1)^{i-k}}{r_i} \prod_{j=k}^{i-1} \frac{s_j}{r_j} a_i M^{-1} \right|^{p'_k} < \infty,$$

$$\sum_{i=k}^{\infty} \frac{(-1)^{i-k}}{r_i} \prod_{j=k}^{i-1} \frac{s_j}{r_j} a_i < \infty.$$

This shows that $\{\ell(\widetilde{B}, p)\}^{\beta} = S_3(p) \cap S_4(p)$. □

Theorem 3.2.12. $\{\ell(\widetilde{B}, p)\}^{\gamma} = \begin{cases} S_3(p) & , \quad 1 < p_k \le H < \infty, \\ S_5(p) & , \quad 0 < p_k \le 1 \end{cases}$ *for all* $k \in \mathbb{N}$.

Proof. From Lemma 3.2.7 and (3.13), we obtain that $ax = (a_i x_i) \in bs$ whenever $x = (x_i) \in \ell(\widetilde{B}, p)$ if and only if $Dy \in \ell_\infty$ whenever $y = (y_k) \in \ell(p)$, where $D = (d_{nk})$ is defined by (3.12). Therefore, we obtain from (3.7) and (3.8) that $\{\ell(\widetilde{B}, p)\}^{\gamma} = S_3(p)$ for $1 < p_k$, $\{\ell(\widetilde{B}, p)\}^{\gamma} = S_5(p)$ for $p_k \le 1$. □

3.2.2 Matrix Transformations on the Sequence Space $\ell(\widetilde{B}, p)$

In this subsection, we characterize some matrix transformations on the space $\ell(\widetilde{B}, p)$. Theorem 3.2.13 gives the exact conditions of the general case $0 < p_k \le H < \infty$ by combining the cases $0 < p_k \le 1$ and $1 < p_k \le H < \infty$. We consider only the case $1 < p_k \le H < \infty$ and leave the case $0 < p_k \le 1$ to the reader because it can be proved in a similar way.

Theorem 3.2.13. *Let $A = (a_{nk})$ be an infinite matrix. Then, the following statements hold:*

(i) Let $1 < p_k \leq H < \infty$ for all $k \in \mathbb{N}$. Then, $A \in (\ell(\widetilde{B}, p) : \ell_\infty)$ if and only if there exists an integer $M > 1$ such that

$$\sup_{n\in\mathbb{N}} \sum_k \left| \sum_{i=k}^{n} \frac{(-1)^{i-k}}{r_i} \prod_{j=k}^{i-1} \frac{s_j}{r_j} a_{ni} M^{-1} \right|^{p_k'} < \infty, \tag{3.14}$$

$$\sum_{i=k}^{\infty} \frac{(-1)^{i-k}}{r_i} \prod_{j=k}^{i-1} \frac{s_j}{r_j} a_{ni} < \infty. \tag{3.15}$$

(ii) Let $0 < p_k \leq 1$ for all $k \in \mathbb{N}$. Then, $A \in (\ell(\widetilde{B}, p) : \ell_\infty)$ if and only if the condition (3.15) holds, and

$$\sup_{n,k\in\mathbb{N}} \left| \sum_{i=k}^{n} \frac{(-1)^{i-k}}{r_i} \prod_{j=k}^{i-1} \frac{s_j}{r_j} a_{ni} \right|^{p_k} < \infty. \tag{3.16}$$

Proof. (i) Suppose that the conditions (3.14) and (3.15) hold, and $x \in \ell(\widetilde{B}, p)$ with $p_k > 1$ for all $k \in \mathbb{N}$. In this situation, since $(a_{nk})_{k\in\mathbb{N}} \in \{\ell(\widetilde{B}, p)\}^\beta$ for every fixed $n \in \mathbb{N}$, the A-transform of x exists. Consider the following equality obtained by using the relation (3.1) that

$$\sum_{k=0}^{m} a_{nk} x_k = \sum_{k=0}^{m} \sum_{i=k}^{m} \frac{(-1)^{i-k}}{r_i} \prod_{j=k}^{i-1} \frac{s_j}{r_j} a_{ni} y_k \tag{3.17}$$

for all $m, n \in \mathbb{N}$. Taking into account the hypothesis we derive from (3.17), as $m \to \infty$, that

$$\sum_k a_{nk} x_k = \sum_k \sum_{i=k}^{\infty} \frac{(-1)^{i-k}}{r_i} \prod_{j=k}^{i-1} \frac{s_j}{r_j} a_{ni} y_k \quad \text{for each} \quad n \in \mathbb{N}. \tag{3.18}$$

Now, by combining (3.18) with the following inequality (see [77]) which holds for any $M > 0$ and any $a, b \in \mathbb{C}$

$$|ab| \leq M \left(|aM^{-1}|^{p'} + |b|^p \right),$$

where $p > 1$ and $p^{-1} + p'^{-1} = 1$, one can easily see that

$$\sup_{n\in\mathbb{N}} \left| \sum_k a_{nk} x_k \right| \leq \sup_{n\in\mathbb{N}} \sum_k \left| \sum_{i=k}^{\infty} \frac{(-1)^{i-k}}{r_i} \prod_{j=k}^{i-1} \frac{s_j}{r_j} a_{ni} \right| |y_k|$$

$$
\begin{aligned}
&\le \sup_{n\in\mathbb{N}} \sum_k M\left(\left| \sum_{i=k}^{\infty} \frac{1}{r_i} \prod_{j=k}^{i-1} \frac{s_j}{r_j} a_{ni} M^{-1} \right|^{p_k'} + |y_k|^{p_k} \right) \\
&\le M \left(\sup_{n\in\mathbb{N}} \sum_k \left| \sum_{i=k}^{\infty} \frac{1}{r_i} \prod_{j=k}^{i-1} \frac{s_j}{r_j} a_{ni} M^{-1} \right|^{p_k'} + \sum_k |y_k|^{p_k} \right) \\
&< \infty.
\end{aligned}
$$

Conversely, suppose that $A \in (\ell(\widetilde{B}, p) : \ell_\infty)$ and $1 < p_k \le H < \infty$ for all $k \in \mathbb{N}$. Then, Ax exists for every $x \in \ell(\widetilde{B}, p)$ and this implies that $(a_{nk})_{k\in\mathbb{N}} \in \{\ell(\widetilde{B}, p)\}^\beta$ for all $n \in \mathbb{N}$. Now, the necessity of (3.15) is immediate. Besides, we have from (3.18) that the matrix $B = (b_{nk})$ defined by

$$
b_{nk} = \sum_{i=k}^{\infty} \frac{(-1)^{i-k}}{r_i} \prod_{j=k}^{i-1} \frac{s_j}{r_j} a_{ni} \text{ for all } n, k \in \mathbb{N}
$$

is in the class $(\ell(p) : \ell_\infty)$. Then, B satisfies the condition (3.8) which is equivalent to (3.14).

This completes the proof. □

Lemma 3.2.14. [108, Theorem 1] $A = (a_{nk}) \in (\ell(p) : f)$ *if and only if (3.7) and (3.8) hold, and*

$$
\exists \alpha_k \in \mathbb{C} \ni \quad f - \lim \; a_{nk} = \alpha_k \; \text{ for every fixed } \; k \in \mathbb{N}. \tag{3.19}
$$

Theorem 3.2.15. *Let the elements of the matrices $E = (e_{nk})$ and $F = (f_{nk})$ be connected with the relation*

$$
e_{nk} := s_{k-1} f_{n,k-1} + r_k f_{nk} \quad or \quad f_{nk} := \sum_{i=k}^{\infty} \frac{(-1)^i}{r_i} \prod_{j=k}^{i-1} \frac{s_j}{r_j} e_{ni}
$$

for all $k, n \in \mathbb{N}$. Then, $E \in (\ell(\widetilde{B}, p) : f)$ if and only if $F \in (\ell(p) : f)$ and

$$
F^n \in (\ell(p) : c)
$$

for every fixed $n \in \mathbb{N}$, where $F^n = \left(f_{mk}^{(n)}\right)$ with

$$
f_{mk}^{(n)} = \begin{cases} \sum_{i=k}^{m} \frac{(-1)^i}{r_i} \prod_{j=k}^{i-1} \frac{s_j}{r_j} e_{ni} &, \quad 0 \le k \le m, \\[2ex] 0 &, \quad k > m \end{cases}
$$

for all $m, k \in \mathbb{N}$.

Proof. Let $E = (e_{nk}) \in (\ell(\widetilde{B},p) : f)$ and take $x \in \ell(\widetilde{B},p)$. Then, we obtain the equality

$$\begin{aligned}\sum_{k=0}^{m} e_{nk}x_k &= \sum_{k=0}^{m} e_{nk}\left[\sum_{i=0}^{k}\frac{(-1)^{k-i}}{r_i}\prod_{j=i}^{k-1}\frac{s_j}{r_j}y_i\right] \\ &= \sum_{k=0}^{m}\left[\sum_{i=k}^{m}\frac{(-1)^{i}}{r_i}\prod_{j=k}^{i-1}\frac{s_j}{r_j}e_{ni}\right]y_k = \sum_{k=0}^{m} f_{mk}^{(n)}y_k\end{aligned} \tag{3.20}$$

for all $m,n \in \mathbb{N}$. Since Ex exists, $F^n \in (\ell(p) : c)$. Letting $m \to \infty$ in the equality (3.20), we have $Ex = Fy$. Since $Ex \in f$, then $Fy \in f$. That is, $F \in (\ell(p) : f)$.

Conversely, let $F \in (\ell(p) : f)$ and $F^n \in (\ell(p) : c)$, and take $x \in \ell(\widetilde{B},p)$. Then, since $(f_{nk})_{k\in\mathbb{N}} \in \{\ell(p)\}^{\beta}$ and $F \in (\ell(p) : f)$ we have $(e_{nk})_{k\in\mathbb{N}} \in \{\ell(\widetilde{B},p)\}^{\beta}$ for all $n \in \mathbb{N}$. So, Ex exists. Therefore, we reobtain from the equality (3.20), as $m \to \infty$, that $Ex = Fy$, that is, $E \in (\ell(\widetilde{B},p) : f)$. □

Theorem 3.2.16. *Let $0 < p_k \le H < \infty$ for all $k \in \mathbb{N}$. Then, $A \in (\ell(\widetilde{B},p) : c)$ if and only if (3.14)-(3.16) hold and*

$$\lim_{n\to\infty}\sum_{k=i}^{\infty}\frac{(-1)^{k-i}}{r_k}\prod_{j=i}^{k-1}\frac{s_j}{r_j}a_{nk} = \alpha_k \quad \textit{for every fixed } \; k \in \mathbb{N}. \tag{3.21}$$

Proof. Let $A \in (\ell(\widetilde{B},p) : c)$ and $1 < p_k \le H < \infty$ for all $k \in \mathbb{N}$. Then, since the inclusion $c \subset \ell_\infty$ holds, the necessities of (3.14) and (3.15) are immediately obtained from Part (i) of Theorem 3.2.13.

To prove the necessity of (3.21), consider the sequence $b^{(k)}$ defined by (3.6) which is in the space $\ell(\widetilde{B},p)$ for every fixed $k \in \mathbb{N}$. Because of the A-transform every $x \in \ell(\widetilde{B},p)$ exists and is in c by the hypothesis,

$$Ab^{(k)} = \left\{\sum_{k=i}^{\infty}\frac{(-1)^{k-i}}{r_k}\prod_{j=i}^{k-1}\frac{s_j}{r_j}a_{nk}\right\}_{n\in\mathbb{N}} \in c$$

for every fixed $k \in \mathbb{N}$ which shows the necessity of (3.21).

Conversely suppose that the conditions (3.14), (3.15) and (3.21) hold, and take any $x = (x_k)$ in the space $\ell(\widetilde{B},p)$. Then, Ax exists. We observe for all $m,n \in \mathbb{N}$ that

$$\begin{aligned}&\sum_{k=0}^{m}\left|\sum_{k=i}^{m}\frac{(-1)^{k-i}}{r_k}\prod_{j=i}^{k-1}\frac{s_j}{r_j}a_{nk}M^{-1}\right|^{p_k'} \\ &\le \sup_{n\in\mathbb{N}}\sum_{k}^{m}\left|\sum_{k=i}^{m}\frac{(-1)^{k-i}}{r_k}\prod_{j=i}^{k-1}\frac{s_j}{r_j}a_{nk}M^{-1}\right|^{p_k'} < \infty\end{aligned}$$

which gives the fact by letting $m,n\to\infty$ with (3.14) and (3.21) that

$$\lim_{m,n\to\infty}\sum_{k=0}^{m}\left|\sum_{k=i}^{m}\frac{(-1)^{k-i}}{r_k}\prod_{j=i}^{k-1}\frac{s_j}{r_j}a_{nk}M^{-1}\right|^{p_k'}$$

$$\leq\sup_{n\in\mathbb{N}}\sum_{k=0}^{m}\left|\sum_{k=i}^{m}\frac{(-1)^{k-i}}{r_k}\prod_{j=i}^{k-1}\frac{s_j}{r_j}a_{nk}M^{-1}\right|^{p_k'}<\infty.$$

This shows that $\sum_k|\alpha_kM^{-1}|^{p_k'}<\infty$. So, $(\alpha_k)_{k\in\mathbb{N}}\in\{\ell(\widetilde{B},p)\}^{\beta}$ which implies that the series $\sum_k\alpha_kx_k$ converges for every $x\in\ell(\widetilde{B},p)$.

Let us now consider the equality obtained from (3.18) with $a_{nk}-\alpha_k$ instead of a_{nk}

$$\begin{aligned}\sum_k(a_{nk}-\alpha_k)x_k &= \sum_i\sum_{k=i}^{m}\frac{(-1)^{k-i}}{r_k}\prod_{j=i}^{k-1}\frac{s_j}{r_j}(a_{nk}-\alpha_k)y_i\\ &= \sum_i c_{ni}y_i \end{aligned}\tag{3.22}$$

for all $n\in\mathbb{N}$, where $C=(c_{ni})$ defined by

$$c_{ni}=\sum_{k=i}^{m}\frac{(-1)^{k-i}}{r_k}\prod_{j=i}^{k-1}\frac{s_j}{r_j}(a_{nk}-\alpha_k)\ \text{ for all }\ m,n,i\in\mathbb{N}.$$

Therefore, we have from Lemma 3.2.8 that the matrix C belongs to the class $(\ell(p):c_0)$ of infinite matrices. Thus, we see by (3.22) that

$$\lim_{n\to\infty}\sum_k(a_{nk}-\alpha_k)x_k=0.\tag{3.23}$$

(3.23) means that $Ax\in c$ whenever $x\in\ell(\widetilde{B},p)$ and this is what we wished to prove. □

Therefore, we have:

Corollary 3.2.17. *Let* $0<p_k\leq H<\infty$ *for all* $k\in\mathbb{N}$. *Then,* $A\in(\ell(\widetilde{B},p):c_0)$ *if and only if (3.14)-(3.16) hold, and (3.21) also holds with* $\alpha_k=0$ *for all* $k\in\mathbb{N}$.

Now, combining Lemma 2.2.30 with Theorems 3.2.13, 3.2.15, 3.2.16 and Corollary 3.2.17, one can derive the following results:

Corollary 3.2.18. *Let* $A=(a_{nk})$ *be an infinite matrix and define the matrix* $C=(c_{nk})$ *by*

$$c_{nk}=\sum_{j=0}^{n}\binom{n}{j}(1-t)^{n-j}t^{j}a_{jk}\ \textit{ for all }\ n,k\in\mathbb{N}.$$

Then, the necessary and sufficient conditions in order for A to belong to any one of the classes $(\ell(\widetilde{B},p):e^t_\infty)$, $(\ell(\widetilde{B},p):e^t_c)$ and $(\ell(\widetilde{B},p):e^t_0)$ are obtained from the respective ones in Theorems 3.2.13, 3.2.16 and Corollary 3.2.17 by replacing the entries of the matrix A by those of the matrix C, where $0<t<1$.

Corollary 3.2.19. *Let $A=(a_{nk})$ be an infinite matrix and define the matrix $C=(c_{nk})$ by*

$$c_{nk}=sa_{n-1,k}+ra_{nk} \quad \text{for all} \quad n,k\in\mathbb{N}.$$

Then, the necessary and sufficient conditions in order for A to belong to the class $(\ell(\widetilde{B},p):\widehat{f})$ are obtained from Theorem 3.2.15 by replacing the entries of the matrix A by those of the matrix C, where $r,s\in\mathbb{R}\setminus\{0\}$ and $\widehat{f}$ denotes the space of all sequences whose $B(r,s)$-transforms are in the space f and is recently studied by Başar and Kirişçi [29].

Corollary 3.2.20. *Let $A=(a_{nk})$ be an infinite matrix and define the matrix $C=(c_{nk})$ by*

$$c_{nk}=ta_{n-2,k}+sa_{n-1,k}+ra_{nk} \quad \text{for all} \quad n,k\in\mathbb{N}.$$

Then, the necessary and sufficient conditions in order for A to belong to the class $(\ell(\widetilde{B},p):f(B))$ are obtained from Theorem 3.2.15 by replacing the entries of the matrix A by those of the matrix C, where $r,s,t\in\mathbb{R}\setminus\{0\}$ and $f(B)$ denotes the space of all sequences whose $B(r,s,t)$-transforms are in the space f and is recently studied by Sönmez [126].

Corollary 3.2.21. *Let $A=(a_{nk})$ be an infinite matrix and define the matrix $C=(c_{nk})$ by*

$$c_{nk}=\frac{1}{n+1}\sum_{j=0}^{n}a_{jk} \quad \text{for all} \quad n,k\in\mathbb{N}.$$

Then, the necessary and sufficient conditions in order for A to belong to the class $(\ell(\widetilde{B},p):\widetilde{f})$ are obtained from Theorem 3.2.15 by replacing the entries of the matrix A by those of the matrix C, where $\widetilde{f}$ denotes the space of all sequences whose C_1-transforms are in the space f and is recently studied by Kayaduman and Şengönül [69].

Corollary 3.2.22. *Let $A=(a_{nk})$ be an infinite matrix and $t=(t_k)$ be a sequence of positive numbers and define the matrix $C=(c_{nk})$ by*

$$c_{nk}=\frac{1}{T_n}\sum_{j=0}^{n}t_ja_{jk} \quad \text{for all} \quad n,k\in\mathbb{N},$$

where $T_n=\sum_{k=0}^{n}t_k$ for all $n\in\mathbb{N}$. Then, the necessary and sufficient conditions in order for A to belong to any one of the classes $(\ell(\widetilde{B},p):r^t_\infty)$, $(\ell(\widetilde{B},p):r^t_c)$ and $(\ell(\widetilde{B},p):r^t_0)$ are obtained from the respective ones in Theorems 3.2.13, 3.2.16 and Corollary 3.2.17 by replacing the entries of the matrix A by those of the matrix C.

Since the spaces r^t_∞, r^t_c and r^t_0 reduce in the case $t = e$ to the Cesàro sequence spaces X_∞, $\widetilde{c}$ and $\widetilde{c}_0$ of non-absolute type, respectively, Corollary 3.2.22 also includes the characterizations of the classes $(\ell(\widetilde{B},p) : X_\infty)$, $(\ell(\widetilde{B},p) : \widetilde{c})$ and $(\ell(\widetilde{B},p) : \widetilde{c}_0)$, as a special case.

Corollary 3.2.23. *Let $A = (a_{nk})$ be an infinite matrix and define the matrix $C = (c_{nk})$ by $c_{nk} = a_{nk} - a_{n+1,k}$ for all $n, k \in \mathbb{N}$. Then, the necessary and sufficient conditions in order for A to belong to any one of the classes $(\ell(\widetilde{B},p) : \ell_\infty(\Delta))$, $(\ell(\widetilde{B},p) : c(\Delta))$ and $(\ell(\widetilde{B},p) : c_0(\Delta))$ are obtained from the respective ones in Theorems 3.2.13, 3.2.16 and Corollary 3.2.17 by replacing the entries of the matrix A by those of the matrix C.*

Corollary 3.2.24. *Let $A = (a_{nk})$ be an infinite matrix and define the matrix $C = (c_{nk})$ by $c_{nk} = \sum_{j=0}^{n} a_{jk}$ for all $n, k \in \mathbb{N}$. Then the necessary and sufficient conditions in order for A to belong to any one of the classes $(\ell(\widetilde{B},p) : bs)$, $(\ell(\widetilde{B},p) : cs)$ and $(\ell(\widetilde{B},p) : cs_0)$ are obtained from the respective ones in Theorems 3.2.13, 3.2.16 and Corollary 3.2.17 by replacing the entries of the matrix A by those of the matrix C.*

3.2.3 The Rotundity of the Space $\ell(\widetilde{B},p)$

The rotundity of Banach spaces is one of the most important geometric properties in functional analysis. For details, the reader may refer to [39], [47] and [90]. In this subsection, we characterize the rotundity of the space $\ell(\widetilde{B},p)$ and examine its geometric properties.

Definition 3.2.25. *Let $S(X)$ be the unit sphere of a Banach space X. Then a point $x \in S(X)$ is called an extreme point if $2x = y + z$ implies $y = z$ for every $y, z \in S(X)$. A Banach space X is said to be rotund (strictly convex) if every point of $S(X)$ is an extreme point.*

Definition 3.2.26. *A Banach space X is said to have Kadec-Klee property (or property (H)) if every weakly convergent sequence on the unit sphere is convergent in norm.*

Definition 3.2.27. *A Banach space X is said to have*

(i) the Opial property if every sequence (x_n) weakly convergent to $x_0 \in X$ satisfies

$$\liminf_{n\to\infty} \|x_n - x_0\| < \liminf_{n\to\infty} \|x_n + x\|$$

for every $x \in X$ with $x \neq x_0$.

(ii) the uniform Opial property if for each $\epsilon > 0$, there exists an $r > 0$ such that

$$1 + r \leq \liminf_{n\to\infty} \|x_n + x\|$$

for each $x \in X$ *with* $\|x\| \geq \epsilon$ *and each sequence* (x_n) *in* X *such that* $x_n \to 0$ *and* $\liminf_{n\to\infty} \|x_n\| \geq 1$.

Definition 3.2.28. *Let* X *be a real vector space. A functional* $\sigma : X \to [0, \infty)$ *is called a modular if*

(i) $\sigma(x) = 0$ *if and only if* $x = \theta$;

(ii) $\sigma(\alpha x) = \sigma(x)$ *for all scalars* α *with* $|\alpha| = 1$;

(iii) $\sigma(\alpha x + \beta y) \leq \sigma(x) + \sigma(y)$ *for all* $x, y \in X$ *and* $\alpha, \beta \geq 0$ *with* $\alpha + \beta = 1$;

(iv) *The modular* σ *is called convex if* $\sigma(\alpha x + \beta y) \leq \alpha\sigma(x) + \beta\sigma(y)$ *for all* $x, y \in X$ *and* $\alpha, \beta > 0$ *with* $\alpha + \beta = 1$.

A modular σ *on* X *is called*

(a) *right continuous if* $\lim_{\alpha\to 1^+} \sigma(\alpha x) = \sigma(x)$ *for all* $x \in X_\sigma$.

(b) *left continuous if* $\lim_{\alpha\to 1^-} \sigma(\alpha x) = \sigma(x)$ *for all* $x \in X_\sigma$.

(c) *continuous if it is both right and left continuous, where*

$$X_\sigma = \left\{x \in X : \lim_{\alpha\to 0^+} \sigma(\alpha x) = 0\right\}.$$

We define σ_p on $\ell(\widetilde{B}, p)$ by $\sigma_p(x) = \sum_k |s_{k-1}x_{k-1} + r_k x_k|^{p_k}$. If $p_k \geq 1$ for all $k \in \mathbb{N}_1 = \{1, 2, 3, \ldots\}$, by the convexity of the function $t \mapsto |t|^{p_k}$ for each $k \in \mathbb{N}_1$, σ_p is a convex modular on $\ell(\widetilde{B}, p)$.

Proposition 3.2.29. *The modular* σ_p *on* $\ell(\widetilde{B}, p)$ *satisfies the following properties with* $p_k \geq 1$ *for all* $k \in \mathbb{N}$:

(i) *If* $0 < \alpha \leq 1$, *then* $\alpha^M \sigma_p(x/\alpha) \leq \sigma_p(x)$ *and* $\sigma_p(\alpha x) \leq \alpha\sigma_p(x)$.

(ii) *If* $\alpha \geq 1$, *then* $\sigma_p(x) \leq \alpha^M \sigma_p(x/\alpha)$.

(iii) *If* $\alpha \geq 1$, *then* $\sigma_p(x) \geq \alpha\sigma_p(x/\alpha)$.

(iv) *The modular* σ_p *is continuous on the space* $\ell(\widetilde{B}, p)$.

Proof. Consider the modular σ_p on $\ell(\widetilde{B}, p)$.

(i) Let $0 < \alpha \leq 1$, then $\alpha^M/\alpha^{p_k} \leq 1$. So, we have

$$\begin{aligned}
\alpha^M \sigma_p\Big(\frac{x}{\alpha}\Big) &= \alpha^M \sum_k \frac{1}{\alpha^{p_k}} |s_{k-1}x_{k-1} + r_k x_k|^{p_k} \\
&= \sum_k \frac{\alpha^M}{\alpha^{p_k}} |s_{k-1}x_{k-1} + r_k x_k|^{p_k}
\end{aligned}$$

$$\leq \sum_k |s_{k-1}x_{k-1} + r_k x_k|^{p_k} = \sigma_p(x),$$

$$\begin{aligned}\sigma_p(\alpha x) &= \sum_k \alpha^{p_k} |s_{k-1}x_{k-1} + r_k x_k|^{p_k} \\ &\leq \alpha \sum_k |s_{k-1}x_{k-1} + r_k x_k|^{p_k} = \alpha\sigma_p(x).\end{aligned}$$

(ii) Let $\alpha \geq 1$. Then, $\alpha^M/\alpha^{p_k} \geq 1$ for all $p_k \geq 1$. So, we have

$$\sigma_p(x) \leq \frac{\alpha^M}{\alpha^{p_k}} \sigma_p(x) = \alpha^M \sigma_p(\frac{x}{\alpha}).$$

(iii) Let $\alpha \geq 1$. Then, $\alpha/\alpha^{p_k} \leq 1$ for all $p_k \geq 1$. So, we have

$$\begin{aligned}\sigma_p(x) &= \sum_k |s_{k-1}x_{k-1} + r_k x_k|^{p_k} \\ &\geq \sum_k \frac{\alpha}{\alpha^{p_k}} |s_{k-1}x_{k-1} + r_k x_k|^{p_k} = \alpha\sigma_p\left(\frac{x}{\alpha}\right)\end{aligned}$$

which completes the proof of Part (iii).

(iv) By (ii) and (iii), one can immediately see for $\alpha > 1$ that

$$\sigma_p(x) \leq \alpha\sigma_p(x) \leq \sigma_p(\alpha x) \leq \alpha^M \sigma_p(x). \tag{3.24}$$

By passing to limit as $\alpha \to 1^+$ in (3.24), we have

$$\lim_{\alpha \to 1^+} \sigma_p(\alpha x) = \sigma_p(x).$$

Hence, σ_p is right continuous. If $0 < \alpha < 1$, by (i) we have

$$\alpha^M \sigma_p(x) \leq \sigma_p(\alpha x) \leq \alpha\sigma_p(x). \tag{3.25}$$

By letting $\alpha \to 1^-$ in (3.25), we observe that $\lim_{\alpha \to 1^-} \sigma_p(\alpha x) = \sigma_p(x)$. Hence, σ_p is also left continuous and so it is continuous.

□

Proposition 3.2.30. *For any $x \in \ell(\widetilde{B}, p)$, the following statements hold:*

(i) If $\|x\| < 1$, then $\sigma_p(x) \leq \|x\|$.

(ii) If $\|x\| > 1$, then $\sigma_p(x) \geq \|x\|$.

(iii) $\|x\| = 1$ if and only if $\sigma_p(x) = 1$.

(iv) $\|x\| < 1$ if and only if $\sigma_p(x) < 1$.

(v) $\|x\| > 1$ if and only if $\sigma_p(x) > 1$.

Proof. Let $x \in \ell(\widetilde{B}, p)$.

(i) Let $\epsilon > 0$ be such that $0 < \epsilon < 1 - \|x\|$. By the definition of $\|\cdot\|$, there exists an $\alpha > 0$ such that $\|x\| + \epsilon > \alpha$ and $\sigma_p(x) \leq 1$. From Parts (i) and (ii) of Proposition 3.2.29, we obtain

$$\sigma_p(x) \leq \sigma_p\big[(\|x\| + \epsilon)\frac{x}{\alpha}\big] \leq (\|x\| + \epsilon)\sigma_p\big(\frac{x}{\alpha}\big) \leq \|x\| + \epsilon.$$

Since ϵ is arbitrary, we have (i).

(ii) If we choose $\epsilon > 0$ such that $0 < \epsilon < 1 - (1/\|x\|)$, then $1 < (1-\epsilon)\|x\| < \|x\|$. By the definition of $\|\cdot\|$ and Part (i) of Proposition 3.2.29, we have

$$1 < \sigma_p\left[\frac{x}{(1-\epsilon)\|x\|}\right] \leq \frac{1}{(1-\epsilon)\|x\|}\sigma_p(x).$$

So $(1-\epsilon)\|x\| < \sigma_p(x)$ for all $\epsilon \in (0, 1 - (1/\|x\|))$. This implies that $\|x\| < \sigma_p(x)$.

(iii) Since σ_p is continuous, by Theorem 1.4 of [90] we directly have (iii).

(iv) This follows from Parts (i) and (iii).

(v) This follows from Parts (ii) and (iii).

□

Now, we consider the space $\ell(\widetilde{B}, p)$ equipped with the Luxemburg norm given by

$$\|x\| = \inf\Big\{\alpha > 0 : \sigma_p\big(\frac{x}{\alpha}\big) \leq 1\Big\}.$$

Theorem 3.2.31. *$\ell(\widetilde{B}, p)$ is a Banach space with Luxemburg norm.*

Proof. Let $S_x = \{\alpha > 0 : \sigma_p(x/\alpha) \leq 1\}$ and $\|x\| = \inf S_x$ for all $x \in \ell(\widetilde{B}, p)$. Then $S_x \subset (0, \infty)$. Therefore, $\|x\| \geq 0$ for all $x \in \ell(\widetilde{B}, p)$.

For $x = \theta$, $\sigma_p(\theta) = 0$ for all $\alpha > 0$. Hence $S_0 = (0, \infty)$ and $\|\theta\| = \inf S_0 = \inf(0, \infty) = 0$.

Let $x \neq \theta$ and $Y = \{kx : k \in \mathbb{C} \text{ and } x \in \ell(\widetilde{B}, p)\}$ be a non-empty subset of $\ell(\widetilde{B}, p)$. Since $Y \not\subseteq S\big[\ell(\widetilde{B}, p)\big]$, there exists $k_1 \in \mathbb{C}$ such that $k_1 x \notin S\big[\ell(\widetilde{B}, p)\big]$. Obviously $k_1 \neq 0$. We assume that $0 < \alpha < 1/k_1$ and $\alpha \in S_x$. Then $(x/\alpha) \in S\big[\ell(\widetilde{B}, p)\big]$. Since $|k_1\alpha| < 1$, we get

$$k_1 x = k_1\alpha\frac{x}{\alpha} \in S\big[\ell(\widetilde{B}, p)\big]$$

which contradicts the assumption. Hence, we obtain that if $\alpha \in S_x$, then $\alpha > 1/|k_1|$. This means that $\|x\| \geq 1/|k_1| > 0$. Thus, we conclude that $\|x\| = 0$ if and only if $x = \theta$.

Now, let $k \neq 0$ and $\alpha \in S_{kx}$. Then, we have

$$\sigma_p\Big(\frac{kx}{\alpha}\Big) \leq 1 \quad \text{and} \quad \frac{kx}{\alpha} \in S\big[\ell\big(\widetilde{B},p\big)\big].$$

Therefore, we obtain

$$\frac{|k|x}{\alpha} = \frac{|k|}{k} \times \frac{kx}{\alpha} \in S\big[\ell\big(\widetilde{B},p\big)\big] \quad \text{and} \quad \frac{\alpha}{|k|} \in S_x.$$

That is, $\|x\| \leq \alpha/|k|$ and $|k|\|x\| \leq \alpha$ for all $\alpha \in S_{kx}$. So, $|k|\|x\| \leq \|kx\|$.

If we take $1/k$ and kx instead of k and x, respectively, then we obtain that

$$\Big|\frac{1}{k}\Big|\|kx\| \leq \Big\|\frac{1}{k}kx\Big\| = \|x\| \quad \text{and} \quad \|kx\| \leq |k|\|x\|.$$

Hence, we get $\|kx\| = |k|\|x\|$. This also holds when $k = 0$.

To prove the triangle inequality, let $x, y \in \ell\big(\widetilde{B},p\big)$ and $\epsilon > 0$ be given. Then, there exist $\alpha \in S_x$ and $\beta \in S_y$ such that $\alpha < \|x\| + \epsilon$ and $\beta < \|y\| + \epsilon$. Since $S\big[\ell\big(\widetilde{B},p\big)\big]$ is convex,

$$\begin{aligned} &\frac{x}{\alpha} \in S\big[\ell\big(\widetilde{B},p\big)\big], \ \frac{y}{\beta} \in S\big[\ell\big(\widetilde{B},p\big)\big], \\ &\frac{(x+y)}{\alpha+\beta} = \frac{\alpha}{\alpha+\beta}\Big(\frac{x}{\alpha}\Big) + \frac{\beta}{\alpha+\beta}\Big(\frac{y}{\beta}\Big) \in S\big[\ell\big(\widetilde{B},p\big)\big]. \end{aligned}$$

Therefore, $\alpha + \beta \in S_{x+y}$. Then, we have $\|x + y\| \leq \alpha + \beta < \|x\| + \|y\| + 2\epsilon$. Since $\epsilon > 0$ was arbitrary, we obtain $\|x + y\| \leq \|x\| + \|y\|$. Hence, $\|x\| = \inf\{\alpha > 0 : \sigma_p(x/\alpha) \leq 1\}$ is a norm on $\ell\big(\widetilde{B},p\big)$.

Now, we need to show that every Cauchy sequence in $\ell\big(\widetilde{B},p\big)$ is convergent according to the Luxemburg norm. Let $\{x_k^{(n)}\}$ be a Cauchy sequence in $\ell\big(\widetilde{B},p\big)$ and $\epsilon \in (0,1)$. Thus, there exists n_0 such that $\|x^{(n)} - x^{(m)}\| < \epsilon$ for all $n, m \geq n_0$. By Part (i) of Proposition 3.2.30, we have

$$\sigma_p\big(x^{(n)} - x^{(m)}\big) \leq \big\|x^{(n)} - x^{(m)}\big\| < \epsilon \tag{3.26}$$

for all $n, m \geq n_0$. This implies that

$$\sum_k \big|\big[\widetilde{B}\big(x^{(n)} - x^{(m)}\big)\big]_k\big|^{p_k} < \epsilon. \tag{3.27}$$

Then, for each fixed k and for all $n, m \geq n_0$,

$$\big|\big[\widetilde{B}\big(x^{(n)} - x^{(m)}\big)\big]_k\big| = \big|\big(\widetilde{B}x^{(n)}\big)_k - \big(\widetilde{B}x^{(m)}\big)_k\big| < \epsilon.$$

Hence, the sequence $\{(\widetilde{B}x^{(n)})_k\}$ is a Cauchy sequence in $\mathbb{R}$. Since $\mathbb{R}$ is complete, there is a $(\widetilde{B}x)_k \in \mathbb{R}$ such that $(\widetilde{B}x^{(m)})_k \to (\widetilde{B}x)_k$ as $m \to \infty$. Therefore, as $m \to \infty$ by (3.27) we have

$$\sum_k \left|\left[\widetilde{B}(x^{(n)} - x)\right]_k\right|^{p_k} < \epsilon$$

for all $n \geq n_0$.

Now, we have to show that (x_k) is an element of $\ell(\widetilde{B}, p)$. Since $(\widetilde{B}x^{(m)})_k \to (\widetilde{B}x)_k$, as $m \to \infty$, we have

$$\lim_{m\to\infty} \sigma_p(x^{(n)} - x^{(m)}) = \sigma_p(x^{(n)} - x).$$

Then, we see by (3.26) that $\sigma_p(x^{(n)} - x) \leq \|x^{(n)} - x\| < \epsilon$ for all $n \geq n_0$. This implies that $x^n \to x$, as $n \to \infty$. So, we have $x = x^{(n)} - (x^{(n)} - x) \in \ell(\widetilde{B}, p)$. Therefore, the sequence space $\ell(\widetilde{B}, p)$ is complete with respect to Luxemburg norm.

This completes the proof. □

Theorem 3.2.32. *The space $\ell(\widetilde{B}, p)$ is rotund if and only if $p_k > 1$ for all $k \in \mathbb{N}$.*

Proof. Let $\ell(\widetilde{B}, p)$ be rotund and choose $k \in \mathbb{N}$ such that $p_k = 1$ for $k < 3$. Consider the following sequences x and y given by

$$x = \left(0, \frac{1}{r_1}, \frac{-s_1}{r_1 r_2}, \frac{s_1 s_2}{r_1 r_2 r_3}, \ldots\right) \text{ and } y = \left(0, 0, \frac{1}{r_2}, \frac{-s_2}{r_2 r_3}, \frac{s_2 s_3}{r_2 r_3 r_4}, \ldots\right).$$

Then, obviously $x \neq y$ and

$$\sigma_p(x) = \sigma_p(y) = \sigma_p\left(\frac{x+y}{2}\right) = 1.$$

By Part (iii) of Proposition 3.2.30, $x, y, (x+y)/2 \in S\left[\ell(\widetilde{B}, p)\right]$ which leads us to the contradiction that the sequence space $\ell(\widetilde{B}, p)$ is not rotund. Hence, $p_k > 1$ for all $k \in \mathbb{N}$.

Conversely, let $x \in S\left[\ell(\widetilde{B}, p)\right]$ and $v, z \in S\left[\ell(\widetilde{B}, p)\right]$ with $x = (v+z)/2$. By convexity of σ_p and Part (iii) of Proposition 3.2.30, we have

$$1 = \sigma_p(x) \leq \frac{\sigma_p(v) + \sigma_p(z)}{2} \leq \frac{1}{2} + \frac{1}{2} = 1$$

which gives that $\sigma_p(v) = \sigma_p(z) = 1$ and

$$\sigma_p(x) = \frac{\sigma_p(v) + \sigma_p(z)}{2}. \tag{3.28}$$

Also we obtain from (3.28) that

$$\sum_k |s_{k-1}x_{k-1} + r_k x_k|^{p_k} = \frac{1}{2}\left(\sum_k |s_{k-1}v_{k-1} + r_k v_k|^{p_k} + \sum_k |s_{k-1}z_{k-1} + r_k z_k|^{p_k}\right).$$

Since $x = (v + z)/2$, we have

$$\begin{aligned}&\sum_k |s_{k-1}(v_{k-1} + z_{k-1}) + r_k(v_k + z_k)|^{p_k} \\ &= \frac{1}{2}\left(\sum_k |s_{k-1}v_{k-1} + r_k v_k|^{p_k} + \sum_k |s_{k-1}z_{k-1} + r_k z_k|^{p_k}\right).\end{aligned}$$

This implies that

$$\begin{aligned}&|s_{k-1}(v_{k-1} + z_{k-1}) + r_k(v_k + z_k)|^{p_k} \\ &= \frac{1}{2}|s_{k-1}v_{k-1} + r_k v_k|^{p_k} + \frac{1}{2}|s_{k-1}z_{k-1} + r_k z_k|^{p_k}\end{aligned} \tag{3.29}$$

for all $k \in \mathbb{N}$. Since the function $t \mapsto |t|^{p_k}$ is strictly convex for all $k \in \mathbb{N}$, it follows by (3.29) that $v_k = z_k$ for all $k \in \mathbb{N}$. Hence, $v = z$. That is, the sequence space $\ell(\widetilde{B}, p)$ is rotund. □

Theorem 3.2.33. *Let $x \in \ell(\widetilde{B}, p)$. Then, the following statements hold:*

(i) $0 < \alpha < 1$ and $\|x\| > \alpha$ imply $\sigma_p(x) > \alpha^M$.

(ii) $\alpha \geq 1$ and $\|x\| < \alpha$ imply $\sigma_p(x) < \alpha^M$.

Proof. Let $x \in \ell(\widetilde{B}, p)$.

(i) Suppose that $\|x\| > \alpha$ with $0 < \alpha < 1$. Then, $\|x/\alpha\| > 1$. By Part (ii) of Proposition 3.2.30, $\|x/\alpha\| > 1$ implies $\sigma_p(x/\alpha) \geq \|x/\alpha\| > 1$. That is, $\sigma_p(x/\alpha) > 1$. Since $0 < \alpha < 1$, by Part (i) of Proposition 3.2.29, we get $\alpha^M \sigma_p(x/\alpha) \leq \sigma_p(x)$. Thus, we have $\alpha^M < \sigma_p(x)$.

(ii) Let $\|x\| < \alpha$ and $\alpha \geq 1$. Then, $\|x/\alpha\| < 1$. By Part (i) of Proposition 3.2.30, $\|x/\alpha\| < 1$ implies $\sigma_p(x/\alpha) \leq \|x/\alpha\| < 1$. That is, $\sigma_p(x/\alpha) < 1$. If $\alpha = 1$, then $\sigma_p(x/\alpha) = \sigma_p(x) < 1 = \alpha^M$. If $\alpha > 1$, then by Part (ii) of Proposition 3.2.29, we have $\sigma_p(x) \leq \alpha^M \sigma_p(x/\alpha)$. This means that $\sigma_p(x) < \alpha^M$.

□

Theorem 3.2.34. *Let (x_n) be a sequence in $\ell(\widetilde{B}, p)$. Then, the following statements hold:*

(i) $\lim_{n\to\infty} \|x_n\| = 1$ implies $\lim_{n\to\infty} \sigma_p(x_n) = 1$.

(ii) $\lim_{n\to\infty} \sigma_p(x_n) = 0$ implies $\lim_{n\to\infty} \|x_n\| = 0$.

Proof. Let (x_n) be a sequence in $\ell(\widetilde{B},p)$.

(i) Let $\lim_{n\to\infty}\|x_n\| = 1$ and $\epsilon \in (0,1)$. Then, there exists $n_0 \in \mathbb{N}$ such that $1-\epsilon < \|x_n\| < \epsilon + 1$ for all $n \geq n_0$. By Parts (i) and (ii) of Theorem 3.2.33, $1-\epsilon < \|x_n\|$ implies $\sigma_p(x_n) > (1-\epsilon)^M$ and $\|x_n\| < \epsilon+1$ implies $\sigma_p(x_n) < (1+\epsilon)^M$ for all $n \geq n_0$. This means $\epsilon \in (0,1)$ and for all $n \geq n_0$ there exists $n_0 \in \mathbb{N}$ such that $(1-\epsilon)^M < \sigma_p(x_n) < (1+\epsilon)^M$. That is, $\lim_{n\to\infty}\sigma_p(x_n) = 1$.

(ii) We assume that $\lim_{n\to\infty}\|x_n\| \neq 0$ and $\epsilon \in (0,1)$. Then, there exists a subsequence (x_{n_k}) of (x_n) such that $\|x_{n_k}\| > \epsilon$ for all $k \in \mathbb{N}$. By Part (i) of Theorem 3.2.33, $0 < \epsilon < 1$ and $\|x_{n_k}\| > \epsilon$ imply $\sigma_p(x_{n_k}) > \epsilon^M$. Thus, $\lim_{n\to\infty}\sigma_p(x_n) \neq 0$ for all $k \in \mathbb{N}$. Hence, we obtain that $\lim_{n\to\infty}\sigma_p(x_n) = 0$ implies $\lim_{n\to\infty}\|x_n\| = 0$.

□

Theorem 3.2.35. *Let $x \in \ell(\widetilde{B},p)$ and $(x^{(n)}) \subset \ell(\widetilde{B},p)$. If $\sigma_p(x^{(n)}) \to \sigma_p(x)$, as $n \to \infty$ and $x_k^{(n)} \to x_k$, as $n \to \infty$ for all $k \in \mathbb{N}$, then $x^{(n)} \to x$, as $n \to \infty$.*

Proof. Let $\epsilon > 0$ be given. Since $\sigma_p(x) = \sum_k \left|\left(\widetilde{B}x\right)_k\right|^{p_k} < \infty$, there exists $k_0 \in \mathbb{N}$ such that

$$\sum_{k=k_0+1}^{\infty} \left|\left(\widetilde{B}x\right)_k\right|^{p_k} < \frac{\epsilon}{3(2^{M+1})}. \tag{3.30}$$

It follows from the fact that

$$\lim_{n\to\infty}\left[\sigma_p(x^{(n)}) - \sum_{k=1}^{k_0}\left|\left(\widetilde{B}x^{(n)}\right)_k\right|^{p_k}\right] = \sigma_p(x) - \sum_{k=1}^{k_0}\left|\left(\widetilde{B}x\right)_k\right|^{p_k}$$

there exists $n_0 \in \mathbb{N}$ such that for all $n \geq n_0$ and for all $k \in \mathbb{N}$

$$\sigma_p(x_{n_k}) - \sum_{k=1}^{k_0}\left|\left(\widetilde{B}x^{(n)}\right)_k\right|^{p_k} < \sigma_p(x) - \sum_{k=1}^{k_0}\left|\left(\widetilde{B}x\right)_k\right|^{p_k} + \frac{\epsilon}{3(2^M)} \tag{3.31}$$

and for all $n \geq n_0$

$$\sum_{k=1}^{k_0}\left|\left\{\widetilde{B}(x^{(n)}-x)\right\}_k\right|^{p_k} < \frac{\epsilon}{3}. \tag{3.32}$$

Therefore, we obtain from (3.30), (3.31) and (3.32) that

$$\begin{aligned}
\sigma_p(x_n - x) &= \sum_{k=1}^{\infty}\left|\left\{\widetilde{B}(x^{(n)}-x)\right\}_k\right|^{p_k} \\
&< \sum_{k=1}^{k_0}\left|\left\{\widetilde{B}(x^{(n)}-x)\right\}_k\right|^{p_k} + \sum_{k=k_0+1}^{\infty}\left|\left\{\widetilde{B}(x^{(n)}-x)\right\}_k\right|^{p_k}
\end{aligned}$$

$$
\begin{aligned}
&< \frac{\epsilon}{3} + 2^M \left[\sum_{k=k_0+1}^{\infty} \left|\left(\widetilde{B}x^{(n)}\right)_k\right|^{p_k} + \sum_{k=k_0+1}^{\infty} \left|\left(\widetilde{B}x\right)_k\right|^{p_k} \right] \\
&< \frac{\epsilon}{3} + 2^M \left[\sigma_p(x_n) - \sum_{k=1}^{k_0} \left|\left(\widetilde{B}x^{(n)}\right)_k\right|^{p_k} + \sum_{k_0+1}^{\infty} \left|\left(\widetilde{B}x\right)_k\right|^{p_k} \right] \\
&< \frac{\epsilon}{3} + 2^M \left[\sigma_p(x) - \sum_{k=1}^{k_0} \left|\left(\widetilde{B}x\right)_k\right|^{p_k} + \frac{\epsilon}{3(2^M)} + \sum_{k=k_0+1}^{\infty} \left|\left(\widetilde{B}x\right)_k\right|^{p_k} \right] \\
&< \frac{\epsilon}{3} + 2^M \left[2 \sum_{k=k_0+1}^{\infty} \left|\left(\widetilde{B}x\right)_k\right|^{p_k} + \frac{\epsilon}{3(2^M)} \right] \\
&< \frac{\epsilon}{3} + 2^M \left[2\frac{\epsilon}{3(2^{M+1})} + \frac{\epsilon}{3(2^M)} \right] = \epsilon.
\end{aligned}
$$

This means that $\sigma_p\left(x^{(n)} - x\right) \to 0$, as $n \to \infty$. By Part (ii) of Theorem 3.2.34, $\sigma_p\left(x^{(n)} - x\right) \to 0$, as $n \to \infty$ implies $\|x_n - x\| \to 0$, as $n \to \infty$. Hence, $x_n \to x$, as $n \to \infty$. □

Theorem 3.2.36. *The sequence space $\ell(\widetilde{B}, p)$ has the Kadec-Klee property.*

Proof. Let $x \in S\left[\ell(\widetilde{B}, p)\right]$ and $\left(x^{(n)}\right) \subset \ell(\widetilde{B}, p)$ such that $\|x^{(n)}\| \to 1$ and $x^{(n)} \xrightarrow{w} x$ be given. By Part (ii) of Theorem 3.2.34, we have $\sigma_p(x^{(n)}) \to 1$ as $n \to \infty$. Also $x \in S\left[\ell(\widetilde{B}, p)\right]$ implies $\|x\| = 1$. By Part (iii) of Proposition 3.2.30, we obtain $\sigma_p(x) = 1$. Therefore, we have $\sigma_p\left(x^{(n)}\right) \to \sigma_p(x)$ as $n \to \infty$.

Since $x^{(n)} \xrightarrow{w} x$ and $q_k : \ell(\widetilde{B}, p) \to \mathbb{R}$ defined by $q_k(x) = x_k$ is continuous, $x_k^{(n)} \to x_k$ as $n \to \infty$ for all $k \in \mathbb{N}$. Therefore, $x^{(n)} \to x$ as $n \to \infty$.

Because any weakly convergent sequence in $\ell(\widetilde{B}, p)$ is convergent, the sequence space $\ell(\widetilde{B}, p)$ has the Kadec-Klee property. □

Theorem 3.2.37. *For any $1 < p < \infty$, the space $(\ell_p)_{\widetilde{B}}$ has the uniform Opial property.*

Proof. Let $\epsilon > 0$ and $\epsilon_0 \in (0, \epsilon)$ be given such that $1 + (\epsilon^p/2) > (1 + \epsilon_0)^p$. Also let $x \in (\ell_p)_{\widetilde{B}}$ and $\|x\| \geq \epsilon$. There exists $k_1 \in \mathbb{N}$ such that

$$\sum_{k=k_1+1}^{\infty} \left|\left(\widetilde{B}x\right)_k\right|^p < \left(\frac{\epsilon_0}{4}\right)^p.$$

Hence, we have

$$\left\| \sum_{k=k_1+1}^{\infty} x_k e_k \right\| < \frac{\epsilon_0}{4}.$$

Furthermore, we have

$$
\begin{aligned}
\epsilon^p &\leq \sum_{k=1}^{k_1} \left|\left(\widetilde{B}x\right)_k\right|^p + \sum_{k=k_1+1}^{\infty} \left|\left(\widetilde{B}x\right)_k\right|^p \\
&< \sum_{k=1}^{k_1} \left|\left(\widetilde{B}x\right)_k\right|^p + \left(\frac{\epsilon_0}{4}\right)^p \\
&< \sum_{k=1}^{k_1} \left|\left(\widetilde{B}x\right)_k\right|^p + \frac{\epsilon^p}{4}
\end{aligned}
$$

which yields that

$$
\frac{3\epsilon^p}{4} < \sum_{k=1}^{k_1} \left|\left(\widetilde{B}x\right)_k\right|^p.
$$

For any weakly null sequence $\left(x^{(m)}\right) \subset S\left[(\ell_p)_{\widetilde{B}}\right]$, since $x_k^{(m)} \to 0$, as $m \to \infty$ for each $k \in \mathbb{N}$, there exists $m_0 \in \mathbb{N}$ such that for all $m > m_0$,

$$
\left\|\sum_{k=1}^{k_1} x_k^{(m)} e_k\right\| < \frac{\epsilon^p}{4}.
$$

Therefore, for all $m > m_0$

$$
\begin{aligned}
\|x^{(m)} + x\| &= \left\|\sum_{k=1}^{k_1} \left(x_k^{(m)} + x_k\right) e_k + \sum_{k=k_1+1}^{\infty} \left(x_k^{(m)} + x_k\right) e_k\right\| \\
&\geq \left\|\sum_{k=1}^{k_1} x_k e_k + \sum_{k=k_1+1}^{\infty} x_k^{(m)} e_k\right\| - \left\|\sum_{k=1}^{k_1} x_k^{(m)} e_k\right\| - \left\|\sum_{k=k_1+1}^{\infty} x_k e_k\right\| \\
&\geq \left\|\sum_{k=1}^{k_1} x_k e_k + \sum_{k=k_1+1}^{\infty} x_k^{(m)} e_k\right\| - \frac{\epsilon^p}{4} - \frac{\epsilon^p}{4}.
\end{aligned}
$$

Moreover,

$$
\begin{aligned}
\left\|\sum_{k=1}^{k_1} x_k e_k + \sum_{k=k_1+1}^{\infty} x_k^{(m)} e_k\right\|^p &= \sum_{k=1}^{k_1} \left|\left(\widetilde{B}x\right)_k e_k\right|^p + \sum_{k=k_1+1}^{\infty} \left|\left(\widetilde{B}x^{(m)}\right)_k e_k\right|^p \\
&\geq \frac{3\epsilon^p}{4} + \left(1 - \frac{\epsilon^p}{4}\right) \\
&= 1 + \frac{\epsilon^p}{2} \\
&> (1 + \epsilon_0)^p.
\end{aligned}
$$

Then, we have

$$\begin{aligned}\|x^{(m)} + x\| &\geq \left\| \sum_{k=1}^{k_1} x_k e_k + \sum_{k=k_1+1}^{\infty} x_k^{(m)} e_k \right\| - \frac{\epsilon^p}{2} \\ &\geq 1 + \epsilon_0 - \frac{\epsilon^p}{2} \\ &> 1 + \frac{\epsilon_0^p}{2}.\end{aligned}$$

This means that $(\ell_p)_{\widetilde{B}}$ has the uniform Opial property. □

3.3 Conclusion

The difference spaces $\ell_\infty(\Delta)$, $c(\Delta)$ and $c_0(\Delta)$ were introduced by Kızmaz [70]. We treat more different than Kızmaz and the other researchers following him, and use the technique for obtaining a new sequence space by the domain of a triangle matrix. Following this way, the domain of some triangle matrices in the sequence space $\ell(p)$ was recently studied and obtained certain topological and geometric results by Altay and Başar [4, 8]; Choudhary and Mishra [40]; Başar et al. [26]; Aydın and Başar [17]. Although $bv(e, p) = [\ell(p)]_\nabla$ is investigated, since $B(1, -1) \equiv \nabla$, these results are more general than those of Başar et al. [26]. Also in case $p_k = p$ for all $k \in \mathbb{N}$, the results of the present chapter are reduced to the corresponding results of the recent paper of Kirişçi and Başar [73]. We should note that the difference spaces $\Delta c_0(p)$, $\Delta c(p)$ and $\Delta \ell_\infty(p)$ of Maddox's spaces $c_0(p)$, $c(p)$ and $\ell_\infty(p)$ were worked by Ahmad and Mursaleen [1].

The sequence spaces $bv(u, p)$ and $bv_\infty(u, p)$ of non-absolute type consisting of all sequences $x = (x_k)$ such that $\{u_k(x_k - x_{k-1})\}$ in the Maddox's spaces $\ell(p)$ and $\ell_\infty(p)$ were introduced by Başar et al. [26] and the rotundity of the space $bv(u, p)$ was examined, where $u = (u_k)$ is a sequence such that $u_k \neq 0$ for all $k \in \mathbb{N}$.

The sequence space $a^r(u, p)$ of non-absolute type consisting of all sequences $x = (x_k)$ such that $A^r x = \{\sum_{k=0}^{n} (1 + r^k) x_k / (n + 1)\} \in \ell(p)$ was studied by Aydın and Başar [17], and some results related to the rotundity of the space $a^r(u, p)$ were given.

Quite recently, the sequence space $\widehat{\ell}(p)$ of non-absolute type consisting of all sequences $x = (x_k)$ such that $B(r, s)x = (s x_{k-1} + r x_k) \in \ell(p)$ was defined by Aydın and Başar [14], and emphasized the rotundity of the space $\widehat{\ell}(p)$ together with some related results.

Although the sequence spaces $a^r(u, p)$ and $\ell(\widetilde{B}, p)$ are not comparable, since in the special case $\widetilde{r} = re$ and $\widetilde{s} = se$ the double sequential band matrix $B(\widetilde{r}, \widetilde{s})$ reduces to the generalized difference matrix $B(r, s)$, the space $\ell(\widetilde{B}, p)$

is more general than the space $\widehat{\ell}(p)$. Similarly, the sequence space $\ell(\widetilde{B},p)$ is also reduced to the space $bv(u,p)$ in the case $\widetilde{r}=(u_k)$ and $\widetilde{s}=(-u_k)$. So, the results on the space $\ell(\widetilde{B},p)$ are much more comprehensive than the results on the space $bv(u,p)$. Additionally, the corresponding theorems on the Kadec-Klee property of the space $\ell(\widetilde{B},p)$ and the uniform Opial property of the space $(\ell_p)_{\widetilde{B}}$ were not given by Başar et al. [26] and Aydın and Başar [14] which make the present results significant.

Wang introduced the sequence space $X_{a(p)}$, in [133]. Although the domain of several triangle matrices in the classical sequence spaces ℓ_p, c_0, c and ℓ_∞ and in the Maddox spaces $\ell(p)$, $c_0(p)$, $c(p)$ and $\ell_\infty(p)$ were investigated by researchers, the domain of Nörlund mean neither in a normed sequence space nor in a paranormed sequence space was not studied, and was still an open problem. So, we have interest in the domain of Nörlund mean in the Maddox space $\ell(p)$. Additionally, we emphasize some geometric properties of the space $N^t(p)$. It is obvious that the matrix N^t is not comparable with the matrices E^r, A^r or $B(r,s)$. So, the present results are new.

Consider the matrix $\widetilde{B}=(\widetilde{b}_{nk})$ defined by (2.1) as the composition of the matrices E_1, C and ∇. We should record from now on that one can study the domain of the matrix $\widetilde{B}$ in the Maddox's spaces $c_0(p)$, $c(p)$ and $\ell_\infty(p)$ for introducing some new paranormed spaces as a natural continuation of this chapter.

Chapter 4

Paranormed Nörlund Sequence Spaces

Abstract. *In this chapter, we introduce the paranormed Nörlund sequence spaces $N^t(p)$ and determine its alpha-, beta- and gamma-duals. We characterize the classes of matrix transformations from the space $N^t(p)$ into any given sequence space μ and from a given sequence space μ into the space $N^t(p)$. Finally, we give the necessary and sufficient condition in order the space $N^t(p)$ to be rotund and present some results related to this concept.*

Keywords. *Domain of an infinite matrix, Nörlund means, Schauder basis, alpha-, beta- and gamma-duals, characterization of the matrix transformations, paranormed sequence spaces.*

4.1 Paranormed Nörlund Sequence Spaces

In this section, following Yeşilkayagil and Başar [137], we emphasize the paranormed Nörlund sequence space $N^t(p)$ consisting of the sequences whose N^t-transforms are in the Maddox's space $\ell(p)$.

4.1.1 The Nörlund Sequence Space $N^t(p)$

In this subsection, as the domain of the Nörlund matrix N^t in the space $\ell(p)$, we define the Nörlund sequence space $N^t(p)$ and prove that $N^t(p)$ is linearly isomorphic to the space $\ell(p)$, where $0 < p_k \leq H < \infty$ for all $k \in \mathbb{N}$. Finally, we give the basis for the space $N^t(p)$.

In [4], Başar and Altay defined the sequence space $r^q(p)$ which consists of all sequences whose R^q-transforms are in the space $\ell(p)$, that is

$$r^q(p) = \{\ell(p)\}_{R^q} \text{ and } r^q_p = (\ell_p)_{R^q},$$

where $R^q = (r^q_{nk})$ is the matrix of Riesz mean.

In [133], Wang defined the sequence space $X_{a(p)}$ consisting of all sequences whose N^t-transforms are in the space ℓ_p which is a Banach space with the norm

$$\|x\|_p = \left(\sum_k \left| \frac{1}{T_k} \sum_{j=0}^{k} t_{k-j} x_j \right|^p \right)^{1/p}, \quad \text{where } 1 \le p < \infty.$$

Now, we introduce the Nörlund sequence space $N^t(p)$ defined by

$$N^t(p) := \left\{ x = (x_k) \in \omega : \sum_k \left| \frac{1}{T_k} \sum_{j=0}^{k} t_{k-j} x_j \right|^{p_k} < \infty \right\}$$

with $0 < p_k \le H < \infty$. It is natural that the space $N^t(p)$ can also be defined with the notation of (1.3) that $N^t(p) = \{\ell(p)\}_{N^t}$.

Theorem 4.1.1. *$N^t(p)$ is a complete linear metric space paranormed by g_7 defined by*

$$g_7(x) = \left(\sum_k \left| \frac{1}{T_k} \sum_{j=0}^{k} t_{k-j} x_j \right|^{p_k} \right)^{1/M} \quad \textit{with } 0 < p_k \le H < \infty. \tag{4.1}$$

Proof. Since this can be shown by a routine verification, we omit details. □

Remark 4.1.2. *One can easily see that the absolute property does not hold on the space $N^t(p)$, that is $g_7(x) \ne g_7(|x|)$ for at least one sequence in the space $N^t(p)$ and this says that $N^t(p)$ is a sequence space of non-absolute type, where $|x| = (|x_k|)$.*

Theorem 4.1.3. *The Nörlund sequence space $N^t(p)$ of non-absolute type is linearly paranorm isomorphic to the space $\ell(p)$, where $0 < p_k \le H < \infty$ for all $k \in \mathbb{N}$.*

Proof. To prove the theorem, we should show the existence of a linear bijection between the spaces $N^t(p)$ and $\ell(p)$ for $0 < p_k \le H < \infty$. Consider the transformation T defined, with the notation of (2.1), from $N^t(p)$ to $\ell(p)$ by $x \mapsto y = Tx = N^t x$. The linearity of T is clear. Further, it is trivial that $x = \theta$ whenever $Tx = \theta$ and hence T is injective.

Let us take any $y \in \ell(p)$ and define the sequence $x = (x_k)$ by (2.2). Therefore, we see from (4.1) that

$$\begin{aligned} g_7(x) &= \left(\sum_k \left| \frac{1}{T_k} \sum_{j=0}^{k} t_{k-j} x_j \right|^{p_k} \right)^{1/M} \\ &= \left(\sum_k \left| \frac{1}{T_k} \sum_{j=0}^{k} t_{k-j} \sum_{i=0}^{j} (-1)^{j-i} D_{j-i} T_i y_i \right|^{p_k} \right)^{1/M} \end{aligned}$$

$$= \left(\sum_k |y_k|^{p_k} \right)^{1/M} = g_2(y) < \infty.$$

This means that $x \in N^t(p)$. Consequently, T is surjective and is paranorm preserving. Hence, T is a linear bijection and this leads us to the fact that the spaces $N^t(p)$ and $\ell(p)$ are linearly paranorm isomorphic.

This step completes the proof. □

We determine the basis for the paranormed space $N^t(p)$.

Theorem 4.1.4. *Define the sequence $b^{(k)}(t) = \{b_n^{(k)}(t)\}_{n\in\mathbb{N}}$ of the elements of the space $N^t(p)$ for every fixed $k \in \mathbb{N}$ by*

$$b_n^{(k)}(t) = \begin{cases} (-1)^{n-k} D_{n-k} T_k & , \quad 0 \le k \le n, \\ 0 & , \quad k > n. \end{cases}$$

Then, the sequence $\{b^{(k)}(t)\}_{k\in\mathbb{N}}$ is a basis for the space $N^t(p)$ and any $x \in N^t(p)$ has a unique representation of the form

$$x = \sum_k \lambda_k(t) b^{(k)}(t), \tag{4.2}$$

where $\lambda_k(t) = (N^t x)_k$ for all $k \in \mathbb{N}$ and $0 < p_k \le H < \infty$.

Proof. Let $0 < p_k \le H < \infty$. It is clear that $\{b^{(k)}(t)\} \subset N^t(p)$, since

$$N^t b^{(k)}(t) = e^{(k)} \in \ell(p) \text{ for all } k \in \mathbb{N}. \tag{4.3}$$

Let $x \in N^t(p)$ be given. For every non-negative integer m, we put

$$x^{[m]} = \sum_{k=0}^{m} \lambda_k(t) b^{(k)}(t). \tag{4.4}$$

Then, we obtain by applying N^t to (4.4) with (4.3) that

$$N^t x^{[m]} = \sum_{k=0}^{m} \lambda_k(t) N^t b^{(k)}(t) = \sum_{k=0}^{m} (N^t x)_k e^{(k)}$$

and

$$\left\{ N^t(x - x^{[m]}) \right\}_i = \begin{cases} 0 & , \quad 0 \le i \le m, \\ (N^t x)_i & , \quad i > m, \end{cases}$$

where $i, m \in \mathbb{N}$. Given $\epsilon > 0$, then there is an integer m_0 such that

$$\left[\sum_{i=m+1}^{\infty} |(N^t x)_i|^{p_k} \right]^{1/M} < \epsilon$$

for all $(m+1) \geq m_0$. Hence,

$$g_7\left[N^t\left(x - x^{[m]}\right)\right] = \left[\sum_{i=m+1}^{\infty} |(N^t x)_i|^{p_k}\right]^{1/M} \leq \left[\sum_{i=m_0}^{\infty} |(N^t x)_i|^{p_k}\right]^{1/M} < \epsilon$$

for all $(m+1) \geq m_0$ which proves that $x \in N^t(p)$ is represented as in (4.2).

Let us show the uniqueness of the representation for $x \in N^t(p)$ given by (4.2). Suppose, on the contrary, that there exists a representation $x = \sum_k \mu_k(t) b^{(k)}(t)$. Since the linear transformation T, from $N^t(p)$ to $\ell(p)$, used in the proof of Theorem 4.1.3 is continuous we have at this stage that

$$(N^t x)_n = \sum_k \mu_k(t) \left\{N^t b^{(k)}(t)\right\}_n = \sum_k \mu_k(t) e_n^{(k)} = \mu_n(t)$$

for all $n \in \mathbb{N}$ which contradicts the fact that $(N^t x)_n = \lambda_n(t)$ for all $n \in \mathbb{N}$. Hence, the representation (4.2) of $x \in N^t(p)$ is unique.

This completes the proof. □

4.1.2 The Alpha-, Beta- and Gamma-duals of the Space $N^t(p)$

In this subsection, we determine the alpha-, beta- and gamma-duals of the space $N^t(p)$. Because the case $0 < p_k \leq 1$ may be established in the similar way to the proof of the case $1 < p_k \leq H < \infty$, we omit details of that case and give the proof only for the case $1 < p_k \leq H < \infty$ in Theorems 4.1.5 and 4.1.6, below.

Theorem 4.1.5. *Let $1 < p_k \leq H < \infty$ for every $k \in \mathbb{N}$. Define the sets $D_1(p)$, $D_2(p)$ and $D_3(p)$, as follows:*

$$D_1(p) := \left\{a = (a_k) \in \omega : \sup_{N \in \mathcal{F}} \sum_k \left|\sum_{n \in N} (-1)^{n-k} a_n D_{n-k} T_k B^{-1}\right|^{p_k'} < \infty\right\},$$

$$D_2(p) := \left\{a = (a_k) \in \omega : \sup_{n \in \mathbb{N}} \sum_k \left|\sum_{i=k}^{n} (-1)^{i-k} a_i D_{i-k} T_k B^{-1}\right|^{p_k'} < \infty \right.$$
$$\left. \text{and } \left\{\left(a_n T_n B^{-1}\right)^{p_k'}\right\} \in \ell_\infty\right\},$$

$$D_3(p) = cs.$$

Then, the following statements hold:

(i) $\{N^t(p)\}^\alpha = D_1(p)$.

(ii) $\{N^t(p)\}^\gamma = D_2(p)$.

(iii) $\{N^t(p)\}^\beta = D_2(p) \cap D_3(p)$.

Proof. (i) Let us take $a = (a_k) \in \omega$. We easily derive with (2.1) that

$$a_n x_n = \sum_{k=0}^{n} (-1)^{n-k} a_n D_{n-k} T_k y_k = (By)_n \tag{4.5}$$

for all $n \in \mathbb{N}$, where $B = (b^t_{nk})$ is defined by (2.9). Thus, we observe by combining (4.5) with Part (ii) of Lemma 3.2.9 that $ax = (a_n x_n) \in \ell_1$ whenever $x = (x_k) \in N^t(p)$ if and only if $By \in \ell_1$ whenever $y = (y_k) \in \ell(p)$. This gives the result that $\{N^t(p)\}^\alpha = D_1(p)$.

(ii) Consider the equality

$$\begin{aligned} \sum_{k=0}^{n} a_k x_k &= \sum_{k=0}^{n-1} \sum_{i=k}^{n} (-1)^{i-k} a_i D_{i-k} T_k y_k + a_n T_n y_n \\ &= (Ey)_n \quad \text{for all} \quad n \in \mathbb{N}, \end{aligned} \tag{4.6}$$

where $E = (e_{nk})$ is defined by (2.12). Thus, we deduce from Part (ii) of Lemma 3.2.7 with (4.6) that $ax = (a_n x_n) \in bs$ whenever $x = (x_k) \in N^t(p)$ if and only if $Ey \in \ell_\infty$ whenever $y = (y_k) \in \ell(p)$. Therefore, we obtain from Lemma 3.2.8 that $\{N^t(p)\}^\gamma = D_2(p)$.

(iii) We see from Lemma 3.2.8 that $ax = (a_n x_n) \in cs$ whenever $x = (x_k) \in N^t(p)$ if and only if $Ey \in c$ whenever $y = (y_k) \in \ell(p)$. Therefore, we derive from Lemma 3.2.8 that $\{N^t(p)\}^\beta = D_2(p) \cap D_3(p)$.

This completes the proof. □

Theorem 4.1.6. *Let $0 < p_k \le 1$ for every $k \in \mathbb{N}$. Define the sets $D_4(p)$ and $D_5(p)$ by*

$$D_4(p) := \left\{ (a_k) \in \omega : \sup_{N \in \mathcal{F}} \sup_{k \in \mathbb{N}} \left| \sum_{n \in N} (-1)^{n-k} a_n D_{n-k} T_k \right|^{p_k} < \infty \right\},$$

$$D_5(p) := \left\{ (a_k) \in \omega : \sup_{n,k \in \mathbb{N}} \left| \sum_{i=k}^{n} (-1)^{i-k} a_i D_{i-k} T_k \right|^{p_k} < \infty \text{ and } \{(a_n T_n)^{p_k}\} \in \ell_\infty \right\}.$$

Then, the following statements hold:

(i) $\{N^t(p)\}^\alpha = D_4(p)$.

(ii) $\{N^t(p)\}^\gamma = D_5(p)$.

(iii) $\{N^t(p)\}^\beta = D_3(p) \cap D_5(p)$.

Proof. This is easily obtained by proceeding as in the proof of Theorem 4.1.5, above, by using the first parts of Lemmas 3.2.7, 3.2.9 and 3.2.8 instead of the second parts. So, we omit details. □

4.1.3 Some Matrix Transformations Related to the Sequence Space $N^t(p)$

In the present subsection, we characterize the matrix transformations from the space $N^t(p)$ into any given sequence space μ and from a given sequence space μ into the space $N^t(p)$. Since $\mu_A \cong \mu$ for any triangle A and any sequence space μ, it is trivial that the equivalence "$x \in \mu_A$ if and only if $y = Ax \in \mu$" holds.

Now, we may give the following theorem:

Theorem 4.1.7. *Suppose that the entries of the infinite matrices $A = (a_{nk})$ and $F = (f_{nk})$ are connected with the relation (2.13) and μ be any given sequence space. Then, $A \in (N^t(p) : \mu)$ if and only if $A_n \in \{N^t(p)\}^\beta$ for all $n \in \mathbb{N}$ and $F \in (\ell(p) : \mu)$.*

Proof. Let μ be any given sequence space. Suppose that (2.13) holds between the entries of the matrices $A = (a_{nk})$ and $F = (f_{nk})$, and take into account that the spaces $N^t(p)$ and $\ell(p)$ are linearly paranorm isomorphic.

Let $A \in (N^t(p) : \mu)$ and take any $y \in \ell(p)$. Then,

$$(FN^t)_{nk} = \sum_{j=k}^{n} f_{nj} a^t_{jk} = \sum_{j=k}^{n} \sum_{i=j}^{n} (-1)^{i-j} D_{i-j} a_{ni} T_j \frac{t_{j-k}}{T_j} = a_{nk},$$

i.e., FN^t exists and $A_n \in \{N^t(p)\}^\beta$ which yields that $F_n \in \ell_1$ for each $n \in \mathbb{N}$. Hence, Fy exists and thus

$$\sum_k f_{nk} y_k = \sum_k \sum_{i=k}^{n} (-1)^{i-k} D_{i-k} a_{ni} T_k \left(\frac{1}{T_k} \sum_{j=0}^{k} t_{k-j} x_j \right) = \sum_k a_{nk} x_k$$

for all $n \in \mathbb{N}$. So, we have that $Fy = Ax$ which leads us to the consequence $F \in (\ell(p) : \mu)$.

Conversely, let $A_n \in \{N^t(p)\}^\beta$ for each $n \in \mathbb{N}$ and $F \in (\ell(p) : \mu)$, and take $x = (x_k) \in N^t(p)$. Then, Ax exists. Therefore, we obtain from the equality

$$\sum_k a_{nk} x_k = \sum_k a_{nk} \left[\sum_{i=0}^{k} (-1)^{k-i} D_{k-i} T_i y_i \right] = \sum_k f_{nk} y_k \quad \text{for all} \quad n \in \mathbb{N}$$

that $Ax = Fy$ and this shows that $A \in (N^t(p) : \mu)$.

This completes the proof. □

By changing the roles of the spaces $N^t(p)$ with μ in Theorem 4.1.7, we have:

Theorem 4.1.8. *Suppose that μ be any given sequence space and the entries of the infinite matrices $A = (a_{nk})$ and $G = (g_{nk})$ are connected with the relation (2.13). Then, $A \in (\mu : N^t(p))$ if and only if $G \in (\mu : \ell(p))$.*

Proof. Let $x = (x_k) \in \mu$ and reconsider the equality (2.20). Then, we have $\{N^t(Ax)\}_n = (Gx)_n$ for all $n \in \mathbb{N}$. Therefore, $Ax \in N^t(p)$ whenever $x \in \mu$; $N^t(Ax) = Gx \in \ell(p)$ whenever $x \in \mu$.

This completes the proof. □

4.1.4 The Rotundity of the Space $N^t(p)$

In this subsection, we give the necessary and sufficient condition in order the space $N^t(p)$ to be rotund and present some results related to this concept.

We define σ_p on $N^t(p)$ by

$$\sigma_p(x) = \sum_k \left| \frac{1}{T_k} \sum_{j=0}^{k} t_{k-j} x_j \right|^{p_k}.$$

If $p_k \geq 1$ for all $k \in \mathbb{N}_1$, by the convexity of the function $t \mapsto |t|^{p_k}$ for each $k \in \mathbb{N}$, σ_p is a convex modular on $N^t(p)$. We consider $N^t(p)$ equipped with Luxemburg norm given by

$$\|x\| = \inf\{\alpha > 0 : \sigma_p(x/\alpha) \leq 1\}. \tag{4.7}$$

$N^t(p)$ is a Banach space with the norm (4.7). This can be shown by the similar way used in the proof of Theorem 7 in [112].

We establish some basic properties for the modular σ_p.

Proposition 4.1.9. *The modular σ_p on $N^t(p)$ satisfies the following properties with $p_k \geq 1$ for all $k \in \mathbb{N}$:*

(i) If $0 < \alpha \leq 1$, then $\alpha^M \sigma_p(x/\alpha) \leq \sigma_p(x)$ and $\sigma_p(\alpha x) \leq \alpha \sigma_p(x)$.

(ii) If $\alpha \geq 1$, then $\sigma_p(x) \leq \alpha^M \sigma_p(x/\alpha)$.

(iii) If $\alpha \geq 1$, then $\sigma_p(x) \geq \alpha \sigma_p(x/\alpha)$.

(iv) The modular σ_p is continuous.

Proof. (i) Let $0 < \alpha \leq 1$. Then $\alpha^M/\alpha^{p_k} \leq 1$ for all $p_k \geq 1$. So, we have

$$\begin{aligned}
\alpha^M \sigma_p(x/\alpha) &= \sum_k \frac{\alpha^M}{\alpha^{p_k}} \left| \frac{1}{T_k} \sum_{j=0}^{k} t_{k-j} x_j \right|^{p_k} \\
&\leq \sum_k \left| \frac{1}{T_k} \sum_{j=0}^{k} t_{k-j} x_j \right|^{p_k} = \sigma_p(x), \\
\sigma_p(\alpha x) &= \sum_k \alpha^{p_k} \left| \frac{1}{T_k} \sum_{j=0}^{k} t_{k-j} x_j \right|^{p_k}
\end{aligned}$$

$$\leq \alpha \sum_k \left| \frac{1}{T_k} \sum_{j=0}^{k} t_{k-j} x_j \right|^{p_k} = \alpha \sigma_p(x).$$

(ii) Let $\alpha \geq 1$. Then $1 \leq \alpha^M / \alpha^{p_k}$ for all $p_k \geq 1$. So, we have

$$\sigma_p(x) \leq \frac{\alpha^M}{\alpha^{p_k}} \sigma_p(x) = \alpha^M \sigma_p(x/\alpha).$$

(iii) Let $\alpha \geq 1$. Then $\alpha / \alpha^{p_k} \leq 1$ for all $p_k \geq 1$. Therefore, one can easily see that

$$\sigma_p(x) = \sum_k \left| \frac{1}{T_k} \sum_{j=0}^{k} t_{k-j} x_j \right|^{p_k} \geq \sum_k \frac{\alpha}{\alpha^{p_k}} \left| \frac{1}{T_k} \sum_{j=0}^{k} t_{k-j} x_j \right|^{p_k} = \alpha \sigma_p \left(\frac{x}{\alpha} \right)$$

which completes the proof of Part (iii).

(iv) If $\alpha > 1$, then we have

$$\begin{aligned} \sum_k \alpha \left| \frac{1}{T_k} \sum_{j=0}^{k} t_{k-j} x_j \right|^{p_k} &\leq \sum_k \alpha^{p_k} \left| \frac{1}{T_k} \sum_{j=0}^{k} t_{k-j} x_j \right|^{p_k} \\ &\leq \sum_k \alpha^M \left| \frac{1}{T_k} \sum_{j=0}^{k} t_{k-j} x_j \right|^{p_k}, \end{aligned}$$

that is

$$\alpha \sigma_p(x) \leq \sigma_p(\alpha x) \leq \alpha^M \sigma_p(x). \tag{4.8}$$

By passing to limit as $\alpha \to 1^+$ in (4.8), we have $\sigma_p(\alpha x) \to \sigma_p(x)$. Hence, σ_p is right continuous.

If $0 < \alpha < 1$, we have

$$\begin{aligned} \sum_k \alpha^M \left| \frac{1}{T_k} \sum_{j=0}^{k} t_{k-j} x_j \right|^{p_k} &\leq \sum_k \alpha^{p_k} \left| \frac{1}{T_k} \sum_{j=0}^{k} t_{k-j} x_j \right|^{p_k} \\ &\leq \sum_k \alpha \left| \frac{1}{T_k} \sum_{j=0}^{k} t_{k-j} x_j \right|^{p_k}, \end{aligned}$$

that is

$$\alpha^M \sigma_p(x) \leq \sigma_p(\alpha x) \leq \alpha \sigma_p(x). \tag{4.9}$$

By letting $\alpha \to 1^-$ in (4.9), we have $\sigma_p(\alpha x) \to \sigma_p(x)$. Hence, σ_p is left continuous.

Since σ_p is both right and left continuous, it is continuous. □

Now, we give some relationships between the modular σ_p and the Luxemburg norm on $N^t(p)$.

Proposition 4.1.10. *For any $x \in N^t(p)$, the following statements hold:*

(i) If $\|x\| < 1$, then $\sigma_p(x) \leq \|x\|$.

(ii) If $\|x\| > 1$, then $\sigma_p(x) \geq \|x\|$.

(iii) $\|x\| = 1$ if and only if $\sigma_p(x) = 1$.

(iv) $\|x\| < 1$ if and only if $\sigma_p(x) < 1$.

(v) $\|x\| > 1$ if and only if $\sigma_p(x) > 1$.

(vi) If $0 < \alpha < 1$ and $\|x\| > \alpha$, then $\sigma_p(x) > \alpha^M$.

(vii) If $\alpha \geq 1$ and $\|x\| < \alpha$, then $\sigma_p(x) < \alpha^M$.

Proof. Let $x \in N^t(p)$.

(i) Let $\epsilon > 0$ such that $0 < \epsilon < 1 - \|x\|$. By the definition of $\|\cdot\|$ in (4.7), there exists an $\alpha > 0$ such that $\|x\| + \epsilon > \alpha$ and $\sigma_p(x/\alpha) \leq 1$. So, we have

$$\begin{aligned} \sigma_p(x) &\leq \sum_k \left(\frac{\|x\| + \epsilon}{\alpha}\right)^{p_k} \left|\frac{1}{T_k}\sum_{j=0}^{k} t_{k-j}x_j\right|^{p_k} \qquad (4.10) \\ &\leq (\|x\| + \epsilon)\sigma_p(x/\alpha) \leq \|x\| + \epsilon. \end{aligned}$$

Since ϵ is arbitrary, we have $\sigma_p(x) \leq \|x\|$ from (4.10).

(ii) If we choose $\epsilon > 0$ such that $0 < \epsilon < 1 - 1/\|x\|$, then $1 < (1-\epsilon)\|x\| < \|x\|$. By the definition of $\|\cdot\|$ in (4.7) and Part (iii) of Proposition 4.1.9, we have

$$1 < \sigma_p\left[\frac{x}{(1-\epsilon)\|x\|}\right] \leq \frac{1}{(1-\epsilon)\|x\|}\sigma_p(x).$$

So, $(1-\epsilon)\|x\| < \|x\|$ for all $\epsilon \in (0, 1 - (1/\|x\|))$. This implies that $\|x\| < \sigma_p(x)$.

(iii) Since σ_p is continuous, by Theorem 1.4 of [77] we directly have (iii).

(iv) This follows from Parts (i) and (iii).

(v) This follows from Parts (ii) and (iii).

(vi) This follows from Part (ii) and Part (i) of Proposition 4.1.9.

(vii) This follows from Part (i) and Part (ii) of Proposition 4.1.9.

□

Theorem 4.1.11. *The space $N^t(p)$ is rotund if and only if $p_k > 1$ for all $k \in \mathbb{N}$.*

Proof. Let $N^t(p)$ be rotund and choose $k \in \mathbb{N}$ such that $p_k = 1$ for all $k < 3$. Consider the sequences x and y given by

$$x = (1, -D_1, D_2, -D_3, D_4, \ldots) \text{ and } y = (0, T_1, -T_1 D_1, T_1 D_2, -T_1 D_3, \ldots).$$

Then, obviously $x \neq y$ and

$$\sigma_p(x) = \sigma_p(y) = \sigma_p\left(\frac{x+y}{2}\right) = 1.$$

By Part (iii) of Proposition 4.1.10, $x, y, (x+y)/2 \in S[N^t(p)]$ which leads us to the contradiction that the sequence space $N^t(p)$ is not rotund. Hence, $p_k > 1$ for all $k \in \mathbb{N}$.

Conversely, let $x \in S[N^t(p)]$ and $v, z \in S[N^t(p)]$ with $x = (v+z)/2$. By convexity of σ_p and Part (iii) of Proposition 4.1.10, we have

$$1 = \sigma_p(x) \leq \frac{\sigma_p(v) + \sigma_p(z)}{2} = 1$$

which gives that

$$\sigma_p(x) = \frac{\sigma_p(v) + \sigma_p(z)}{2}. \tag{4.11}$$

Also, since $x = (v+z)/2$ we obtain from (4.11) that

$$\sum_k \left| \frac{1}{T_k} \sum_{j=0}^{k} t_{k-j} \frac{(v_j + z_j)}{2} \right|^{p_k}$$
$$= \frac{1}{2}\left(\sum_k \left| \frac{1}{T_k} \sum_{j=0}^{k} t_{k-j} v_j \right|^{p_k} + \sum_k \left| \frac{1}{T_k} \sum_{j=0}^{k} t_{k-j} z_j \right|^{p_k} \right).$$

This implies that

$$\left| \frac{v_j + z_j}{2} \right|^{p_k} = \frac{|v_j|^{p_k} + |z_j|^{p_k}}{2} \tag{4.12}$$

for all $k \in \mathbb{N}$. Since the function $t \to |t|^{p_k}$ is strictly convex for all $k \in \mathbb{N}$, it follows by (4.12) that $v_k = z_k$ for all $k \in \mathbb{N}$. Hence, $v = z$. That is, $N^t(p)$ is rotund. □

Theorem 4.1.12. *Let (x_n) be a sequence in $N^t(p)$. Then, the following statements hold:*

(i) $\lim_{n\to\infty} \|x_n\| = 1$ *implies* $\lim_{n\to\infty} \sigma_p(x_n) = 1$.

(ii) $\lim_{n\to\infty} \sigma_p(x_n) = 0$ *implies* $\lim_{n\to\infty} \|x_n\| = 0$.

Proof. The proof is similar to that of Theorem 10 in [112]. So, we omit details. □

Theorem 4.1.13. *Let $x \in N^t(p)$ and $(x^{(n)}) \subset N^t(p)$. If $\sigma_p(x^{(n)}) \to \sigma_p(x)$, as $n \to \infty$ and $x_k^{(n)} \to x_k$, as $n \to \infty$ for all $k \in \mathbb{N}$, then $x^{(n)} \to x$, as $n \to \infty$.*

Proof. Let $\epsilon > 0$ be given. Since $x \in N^t(p)$ and $(x^{(n)}) \subset N^t(p)$, $\sigma_p(x^{(n)} - x) = \sum_k |\{N^t(x^{(n)} - x)\}_k|^{p_k} < \infty$. So, there exists an $k_0 \in \mathbb{N}$ such that

$$\sum_{k=k_0+1}^{\infty} |\{N^t(x^{(n)} - x)\}_k|^{p_k} < \frac{\epsilon}{2}. \tag{4.13}$$

Also, since $x_k^{(n)} \to x_k$, as $n \to \infty$, we have

$$\sum_{k=0}^{k_0} |\{N^t(x^{(n)} - x)\}_k|^{p_k} < \frac{\epsilon}{2}. \tag{4.14}$$

Therefore, we obtain from (4.13) and (4.14) that $\sigma_p(x^{(n)} - x) < \epsilon$. This means that $\sigma_p(x^{(n)} - x) \to 0$, as $n \to \infty$. This result implies $\|x^{(n)} - x\| \to 0$, as $n \to \infty$ from Part (ii) of Theorem 4.1.12. Hence, $x_n \to x$, as $n \to \infty$. □

Theorem 4.1.14. *The sequence space $N^t(p)$ has the Kadec-Klee property.*

Proof. Let $x \in S[N^t(p)]$ and $(x^{(n)}) \subset N^t(p)$ such that $\|x^{(n)}\| \to 1$ and $x^{(n)} \xrightarrow{w} x$ are given. By Part (i) of Theorem 4.1.12, we have $\sigma_p(x^{(n)}) \to 1$, as $n \to \infty$. Also, $x \in S[N^t(p)]$ implies $\|x\| = 1$. By Part (iii) of Proposition 4.1.10, we obtain $\sigma_p(x) = 1$. Therefore, we have $\sigma_p(x^{(n)}) \to \sigma_p(x)$, as $n \to \infty$.

Since $x^{(n)} \xrightarrow{w} x$ and $q_k : N^t(p) \to \mathbb{R}$ (or $\mathbb{C}$) defined by $q_k(x) = x_k$ is continuous, $x_k^{(n)} \to x_k$, as $n \to \infty$. Therefore, $x^{(n)} \to x$, as $n \to \infty$.

This completes the proof. □

Theorem 4.1.15. *For any $1 < p < \infty$, the space $X_{a(p)}$ has the uniform Opial property.*

Proof. Since the proof can be given by the similar way used in proving Theorem 13 of Nergiz and Başar [112], we omit details. □

4.2 Conclusion

It is clear that by depending the choice of the sequence space μ, the characterization of several classes of matrix transformations from the space

$N^t(p)$ and into the space $N^t(p)$ can be obtained from Theorems 4.1.7 and 4.1.8, respectively. As a natural continuation of this chapter, Yeşilkayagil and Başar [138] recently studied the domain of the Nörlund mean in the Maddox's spaces $\ell_\infty(p)$, $c(p)$ and $c_0(p)$.

Although the matrix transformations from Maddox's spaces $c_0(p)$, $c(p)$ and $\ell_\infty(p)$ to the spaces f_0 and f of almost null and almost convergent sequences are characterized by Nanda [109], the matrix transformations from the domain of a matrix in Maddox's spaces to the spaces f_0 and f were not characterized, until now. So, Theorem 3.1.11 has a special importance for this type of studies.

Consider the matrix $\widetilde{B} = (\widetilde{b}_{nk})$ defined by (2.1) as the composition of the matrices E_1, C and ∇. We should record from now on that one can study the domain of the matrix $\widetilde{B}$ in the Maddox's space $\ell(p)$ and Nanda's space $f(p)$ for introducing some new paranormed spaces, as a natural continuation of this chapter.

Chapter 5

Generalized Orlicz Sequence Spaces

Abstract. *In this chapter, we deal with various ways to generalize Orlicz sequence spaces systematically and investigate several structural properties of such spaces.*

Keywords. *Orlicz sequence spaces, Difference sequences, Generalized Orlicz sequence spaces, Modular sequence spaces.*

5.1 Orlicz Sequence Spaces

Orlicz sequence spaces were first taken up for study with a specific purpose in the theory of Banach spaces. Lindberg [75, 76] found interest in Orlicz spaces in connection with finding Banach spaces with symmetric Schauder bases having complementary subspaces isomorphic to c_0 or ℓ_p $(1 \leq p < \infty)$. Then, it was Lindenstrauss and Tzafriri [78, 79, 80], who investigated Orlicz sequence spaces in more detail and solved many important structural problems in Banach spaces. During the same period, Orlicz sequence spaces were generalized to modular sequences spaces by Woo [132]. It is to be noted that Orlicz sequence spaces are the special cases of Orlicz spaces (see, [62]), and Orlicz spaces have found a number of applications in various fields including in the theory of nonlinear integral equations. Also, Orlicz sequence spaces are the generalization of ℓ_p-spaces, and L_p-spaces find themselves enveloped in Orlicz spaces.

A function $M : [0, \infty) \to [0, \infty)$ which is convex with $M(v) \geq 0$ for $v \geq 0$, and $M(v) \to \infty$ as $v \to \infty$, is called an Orlicz function. It is always possible to represent an Orlicz function in the following integral form

$$M(v) = \int_0^v k(x)dx,$$

where k, the kernel of M, is right differentiable for $x \geq 0$, $k(0) = 0$, $k(x) > 0$ for $x > 0$, k is non-decreasing and $k(x) \to \infty$ as $x \to \infty$ whenever $\frac{M(v)}{v} \uparrow \infty$ as $v \uparrow \infty$.

Let M be an Orlicz function with k as its kernel. Let

$$l(y) = \sup\{x : k(x) \leq y\}.$$

Then, l possesses the same properties as the function k. Let us define Φ by the following formula

$$\Phi(t) = \int_0^t l(y)dy.$$

Then, Φ is an Orlicz function, and we call M and Φ as mutually complementary Orlicz functions. Also, we have the following well-known results:

(i) For all $v, t \geq 0$, $vt \leq M(v) + \Phi(t)$ (Young's inequality).

(ii) For all $v \geq 0$, $vk(v) = M(v) + \Phi(k(v))$.

(iii) For all $v \geq 0$ and $0 < \lambda < 1$, $M(\lambda v) < \lambda M(v)$.

An Orlicz function M is said to satisfy the Δ_2-condition for small v or at 0 if for each $n \in \mathbb{N}$, there exists $R_n > 0$ and $v_n > 0$ such that $M(nv) \leq R_n M(v)$ for all $v \in (0, v_n]$. Moreover, an Orlicz function M is said to satisfy the Δ_2-condition if and only if

$$\limsup_{v \to 0^+} \frac{M(2v)}{M(v)} < \infty.$$

Let M_1 and M_2 be two Orlicz functions. Then, they are said to be equivalent if there are positive constants α, β and $a \in \mathbb{R}$ such that

$$M_1(\alpha v) \leq M_2(v) \leq M_1(\beta v)$$

for all $v \in [0, a]$.

Orlicz introduced a sequence space ℓ_M by using an Orlicz function M (see, Musielak [82]; Lindenstrauss and Tzafriri [78]), as follows,

$$\ell_M := \left\{ x = (x_i) \in \omega : \sum_i M\left(\frac{|x_i|}{\rho}\right) < \infty \text{ for some } \rho > 0 \right\}.$$

We refer to [64, Ch. 4, Sec. 8] for more details about Orlicz sequence spaces.

5.2 Orlicz Sequence Spaces Generated by Difference Operator

In this section, we present a systematic approach to construct generalized Orlicz sequence spaces for difference sequences.

The notion of difference sequence space was introduced by Kızmaz [70] and studied the following spaces

$$Z(\Delta) := \{x = (x_i) \in \omega : \Delta x = \{(\Delta x)_i\} \in Z\},\ Z \in \{c_0, c, \ell_\infty\},$$

where $(\Delta x)_i = x_i - x_{i+1}$ and $(\Delta^0 x)_i = x_i$ for all $i \in \mathbb{N}$.

A sequence $x = (x_i)$ is said to be Δ-convergent if the limit $\lim_i x_i$ exists and is finite. It is easy to see that every convergent sequence is Δ-convergent, but the converse may not be true. Indeed if we consider the sequence $x = (x_i)$ defined by

$$x_i = i + 1 \text{ for all } i \in \mathbb{N}$$

then $(\Delta x)_i = x_i - x_{i+1} = -1$ for each $i \in \mathbb{N}$. So, $x = (x_i)$ is divergent, but it is Δ-convergent.

Let M and Φ be mutually complementary Orlicz functions. Consider the following two sets (see also, Dutta [51])

$$\widetilde{\ell}_M(\Delta) := \left\{x \in \omega : \delta_\Delta(M, x) = \sum_i M(|(\Delta x)_i|) < \infty\right\}$$

and

$$\ell_M(\Delta) := \left\{x \in \omega : \sum_i (\Delta x)_i y_i \text{ converges for all } y \in \widetilde{\ell}_\Phi\right\}.$$

The above two sets are generalizations of the sets $\widetilde{\ell}_M$ and ℓ_M (see [64]), respectively. We can write

$$\widetilde{\ell}_M(\Delta^0) = \widetilde{\ell}_M$$

and

$$\ell_M(\Delta^0) = \ell_M,$$

respectively.

It is easy to see that $\widetilde{\ell}_M(\Delta) \subseteq \ell_M(\Delta)$. Following the techniques provided in Kamthan and Gupta [64] (see also, Dutta [51]), the following results can be obtained. However, we will give sketch of proofs in some cases to make the developments easily understandable to readers.

Proposition 5.2.1. *If $x \in \ell_M(\Delta)$, then*

$$\sup\left\{\left|\sum_i (\Delta x)_i y_i\right| : \delta(\Phi, y) \leq 1\right\} < \infty.$$

Proposition 5.2.1 inspires the definition of a norm on $\ell_M(\Delta)$, as follows,

$$\|x\|_M^\Delta = |x_1| + \sup\left\{\left|\sum_i (\Delta x)_i y_i\right| : \delta(\Phi, y) \leq 1\right\},$$

where $x \in \ell_M(\Delta)$.

Theorem 5.2.2. *$(\ell_M(\Delta), \|\cdot\|_M^\Delta)$ is a normed linear space.*

Proof. It is obvious that $\|\theta\|_M^\Delta = 0$. If $\|x\|_M^\Delta = 0$; then it follows that $|x_0| = 0$, i.e.,

$$x_0 = 0 \tag{5.1}$$

and

$$\left|\sum_i (\Delta x)_i y_i\right| = 0 \text{ for all } y \text{ with } \delta(\Phi, y) \leq 1. \tag{5.2}$$

Taking $y = \{e^{(i)}\}$ if $\Phi(1) \leq 1$, otherwise $y = \left\{\frac{e^{(i)}}{\Phi(1)}\right\}$ in (5.2), we get

$$(\Delta x)_i = 0 \text{ for all } i \in \mathbb{N}. \tag{5.3}$$

Considering $i = 0$ in (5.3) and using (5.1), we get $x_1 = 0$. Proceeding similarly, we get $x_i = 0$ for all $i \in \mathbb{N}$.

It is easy to see that

$$\|\alpha x\|_M^\Delta = |\alpha| \|x\|_M^\Delta \text{ for any scalar } \alpha \text{ and } x \in \ell_M(\Delta)$$

and

$$\|x + y\|_M^\Delta \leq \|x\|_M^\Delta + \|y\|_M^\Delta \text{ for } x, y \in \ell_M(\Delta).$$

□

Theorem 5.2.3. *The space $\ell_M(\Delta)$ is a Banach space under the norm $\|\cdot\|_M^\Delta$.*

Proof. Let (x^i) be a Cauchy sequence in $\ell_M(\Delta)$, where $x^i = (x_n^i) = (x_0^i, x_1^i, \ldots) \in \ell_M(\Delta)$ for all $i \in \mathbb{N}$. Then, for each $\epsilon > 0$, there exists a positive integer n_0 such that

$$\|x^i - x^j\|_M^\Delta < \epsilon \text{ for all } i, j \geq n_0.$$

It follows from the definition of norm and completeness of $\mathbb{C}$ that for all $i, j \geq n_0$

$$\lim_{i\to\infty} x_0^i = l_1 \text{ (say), exists}$$

and also

$$\left|\sum_n \left\{\Delta\left(x^i - x^j\right)\right\}_n y_n\right| < \epsilon$$

for all y with $\delta(\Phi, y) \leq 1$.

It follows from the second part on considering y similarly as used in the proof of Theorem 5.2.2 and using the completeness of $\mathbb{C}$ that

$$\lim_{i\to\infty} (\Delta x^i)_n = y_n \text{ (say), exists,}$$

for all $n \in \mathbb{N}$. Thus,

$$\lim_{i\to\infty} x_n^i = x_n \text{ (say), exists for all } n \in \mathbb{N}.$$

Keeping $i \geq n_0$ fixed and letting $j \to \infty$, it also follows that

$$(x^i - x) \in \ell_M(\Delta),$$

where $x = (x_n)$. Then, using the linearity of $\ell_M(\Delta)$, we can have

$$x \in \ell_M(\Delta).$$

□

The above proof also makes it easy to see that $\|x^i\| \to 0$ implies that $x_n^i \to 0$ for each $i \in \mathbb{N}$. So, we can state the following result:

Theorem 5.2.4. *The space $\ell_M(\Delta)$ is a BK-space under the norm $\|\cdot\|_M^\Delta$.*

Next, we will see that the space $\ell_M(\Delta)$ can be made a normed space under a different norm defined, as follows

$$\|x\|_{(M)}^\Delta = |x_1| + \inf\left\{\rho > 0 : \sum_i M\left(\frac{|(\Delta x)_i|}{\rho}\right) \leq 1\right\}.$$

Theorem 5.2.5. *The space $\left(\ell_M(\Delta), \|\cdot\|_{(M)}^\Delta\right)$ is a normed linear space.*

Let us consider the backward difference operator ∇ so that for a sequence $x = (x_i)$, we have

$$(\nabla x)_i = x_i - x_{i-1},\ x_{-1} = 0;\ i \in \mathbb{N}.$$

It is obvious that a sequence $x = (x_i)$ is Δ-convergent if and only if it is ∇-convergent.

Then, we can define two norms on $\ell_M(\nabla)$, as follows,

$$\|x\|_M^\nabla = \sup\left\{\left|\sum_i (\nabla x)_i y_i\right| : \delta(\Phi, y) \leq 1\right\}$$

and

$$\|x\|^{\nabla}_{(M)} = \inf\left\{\rho > 0 : \sum_i M\left(\frac{|(\nabla x)_i|}{\rho}\right) \leq 1\right\},$$

where $x = (x_i) \in \ell_M(\nabla)$.

It is obvious that $x \in \ell_M(\Delta)$ if and only if $x \in \ell_M(\nabla)$. Also, the space $\ell_M(\nabla)$ can be made a BK-space under the norm $\|\cdot\|^{\nabla}_M$. In fact, it can be established that both the norms $\|\cdot\|^{\nabla}_M$ and $\|\cdot\|^{\nabla}_{(M)}$ on $\ell_M(\nabla)$ are equivalent.

It is obvious that $\|\theta\|^{\nabla}_{(M)} = 0$. Next, assume that $\|x\|^{\nabla}_{(M)} = 0$. Then for a given $\epsilon > 0$, there exists some $\rho_\epsilon \in (0, \epsilon)$ such that

$$M\left(\frac{|(\nabla x)_i|}{\rho_\epsilon}\right) \leq 1$$

for all $i \in \mathbb{N}$. So,

$$M\left(\frac{|(\nabla x)_i|}{\epsilon}\right) \leq M\left(\frac{|(\nabla x)_i|}{\rho_\epsilon}\right) \leq 1$$

for all $i \in \mathbb{N}$.

Suppose that $(\nabla x)_{n_i} \neq 0$ for some $n_i \in \mathbb{N}$. Then, $\frac{|(\nabla x)_{n_i}|}{\epsilon} \to \infty$ as $\epsilon \to 0$. It follows that $M\left(\frac{|(\nabla x)_{n_i}|}{\epsilon}\right) \to \infty$ as $\epsilon \to 0$ for some $n_i \in \mathbb{N}$. This contraction leads to $(\nabla x)_i = 0$ for all $i \in \mathbb{N}$. If $i = 0$, then $x_0 = 0$. This gives $x_1 = 0$ on taking $i = 1$. Proceeding similarly, we get $x_i = 0$ for all $i \in \mathbb{N}$. Thus, $x = \theta$.

Now we establish the triangle inequality, as follows.

The following result is immediate.

Proposition 5.2.6. *The following inequality*

$$\sum_i M\left(\frac{|(\nabla x)_i|}{\|x\|^{\nabla}_{(M)}}\right) \leq 1$$

holds for $x = (x_i) \in \ell_M(\nabla)$.

Using the convexity and non-decreasing properties of M, we get by using Proposition 5.2.6 for $x, y \in \ell_M(\nabla)$ that

$$\begin{aligned}
&\sum_i M\left(\frac{|\nabla(x_i + y_i)|}{\|x\|^{\nabla}_{(M)} + \|y\|^{\nabla}_{(M)}}\right) = \sum_i M\left(\frac{|(\nabla x)_i + (\nabla y)_i|}{\|x\|^{\nabla}_{(M)} + \|y\|^{\nabla}_{(M)}}\right)\\
&\leq \frac{\|x\|^{\nabla}_{(M)}}{\|x\|^{\nabla}_{(M)} + \|y\|^{\nabla}_{(M)}} \sum_i M\left(\frac{|(\nabla x)_i|}{\|x\|^{\nabla}_{(M)}}\right) + \frac{\|y\|^{\nabla}_{(M)}}{\|x\|^{\nabla}_{(M)} + \|y\|^{\nabla}_{(M)}} \sum_i M\left(\frac{|(\nabla y_i)|}{\|y\|^{\nabla}_{(M)}}\right)\\
&< 1.
\end{aligned}$$

Consequently,

$$\|x+y\|^{\nabla}_{(M)} \leq \|x\|^{\nabla}_{(M)} + \|y\|^{\nabla}_{(M)} \text{ holds.}$$

For any scalar α and choosing $\mu = \frac{\rho}{|\alpha|}$, we get for $x \in \ell_M(\nabla)$,

$$\begin{aligned}\|\alpha x\|^{\nabla}_{(M)} &= \inf\left\{\rho > 0 : \sum_i M\left(\frac{|\alpha(\nabla x)_i|}{\rho}\right) \leq 1\right\} \\ &= \inf\left\{\mu|\alpha| > 0 : \sum_i M\left(\frac{|(\nabla x)_i|}{\mu}\right) \leq 1\right\} \\ &= |\alpha|\|\alpha x\|^{\nabla}_{(M)}.\end{aligned}$$

Using similar arguments as applied in Kamthan and Gupta [64, p. 302] (see also, Dutta [51]), we can establish the following result:

Proposition 5.2.7. *Let $x \in \ell_M(\nabla)$ with $\|x\|^{\nabla}_M \leq 1$. Then $x \in \widetilde{\ell}_M(\nabla)$ and $\delta_\nabla(M,x) \leq \|x\|^{\nabla}_M$.*

Now, the following result is immediate:

Proposition 5.2.8. *The following inequality*

$$\sum_i M\left(\frac{|(\nabla x)_i|}{\|x\|^{\nabla}_M}\right) \leq 1$$

holds for $x = (x_i) \in \ell_M(\nabla)$.

In view of Proposition 5.2.8, the following inequality

$$\|x\|^{\nabla}_{(M)} \leq \|x\|^{\nabla}_M$$

holds for $x \in \ell_M(\nabla)$.

If $\|x\|^{\nabla}_{(M)} \leq 1$ for some $x \in \ell_M(\nabla)$, then $x \in \widetilde{\ell}_M(\nabla)$ and $\delta_\nabla(M,x) \leq 1$. Indeed, using Proposition 5.2.6, we get

$$\frac{1}{\|x\|^{\nabla}_{(M)}} \sum_i M\left(|(\nabla x)_i|\right) \leq \sum_i M\left(\frac{|(\nabla x)_i|}{\|x\|^{\nabla}_{(M)}}\right) \leq 1.$$

Thus,

$$\frac{x}{\|x\|^{\nabla}_{(M)}} \in \widetilde{\ell}_M(\nabla) \text{ with } \delta_\nabla\left(M, \frac{x}{\|x\|^{\nabla}_{(M)}}\right) \leq 1.$$

For an arbitrary $z \in \widetilde{\ell}_M(\nabla)$, it can be seen using Young's inequality that

$$\|z\|^{\nabla}_M = \sup\left\{\left|\sum_i (\nabla z)_i y_i\right| : \delta(\Phi, y) \leq 1\right\} \leq \delta_\nabla(M,z) + 1,$$

by Proposition 5.2.6. So,

$$\left\| \frac{x}{\|x\|_{(M)}^{\nabla}} \right\|_M^{\nabla} \leq \sum_i M\left(\frac{|(\nabla x)_i|}{\|x\|_{(M)}^{\nabla}}\right) + 1 \leq 2$$

for $x \in \ell_M(\nabla)$. Thus,

$$\|x\|_M^{\nabla} \leq 2\|x\|_{(M)}^{\nabla}$$

for $x \in \ell_M(\nabla)$.

Hence, we have the following results:

Theorem 5.2.9. *The norms* $\|\cdot\|_M^{\nabla}$ *and* $\|\cdot\|_{(M)}^{\nabla}$ *on* $\ell_M(\nabla)$ *are equivalent.*

Corollary 5.2.10. *The norms* $\|\cdot\|_M^{\Delta}$ *and* $\|\cdot\|_{(M)}^{\Delta}$ *on* $\ell_M(\Delta)$ *are equivalent.*

Corollary 5.2.11. *Let* A *denote any of the difference operators* Δ *or* ∇. *Then, the space* $\left(\ell_M(A), \|\cdot\|_{(M)}^{\Delta}\right)$ *is a BK-space.*

Using Proposition 5.2.8, it is possible to establish that the class $\ell_M(\nabla)$ can also be defined as

$$\ell_M(\nabla) := \left\{ x \in \omega : \sum_i M\left(\frac{|(\nabla x)_i|}{\rho}\right) < \infty \text{ for some } \rho > 0 \right\}.$$

We now introduce the subspace $h_M(\nabla)$ of $\ell_M(\nabla)$ (see Gribanov [57]) as follows

$$h_M(\nabla) := \left\{ x \in \omega : \sum_i M\left(\frac{|(\nabla x)_i|}{\rho}\right) < \infty \text{ for each } \rho > 0 \right\}.$$

Next, we shall use $\|\cdot\|$ instead of $\|\cdot\|_{(M)}^{\nabla}$ just for simplicity. The topology of $h_M(\nabla)$ is the one it inherits from $\|\cdot\|$.

Let $x \in h_M(\nabla)$. Then, for each $\epsilon \in (0,1)$, we can find an n_0 such that

$$\sum_{i \geq n_0} M\left(\frac{|(\nabla x)_i|}{\epsilon}\right) \leq 1.$$

Hence for $n \geq n_0$, it holds

$$\begin{aligned} \|x - x^{(n)}\| &= \inf\left\{ \rho > 0 : \sum_{i \geq n+1} M\left(\frac{|(\nabla x)_i|}{\rho}\right) \leq 1 \right\} \\ &\leq \inf\left\{ \rho > 0 : \sum_{i \geq n} M\left(\frac{|(\nabla x)_i|}{\rho}\right) \leq 1 \right\} < \epsilon. \end{aligned}$$

Thus, $h_M(\nabla)$ is an AK-space for any Orlicz function M.

Now, we show that $h_M(\nabla)$ is a BK-space. It is enough to establish that $h_M(\nabla)$ is a closed subspace of $\ell_M(\nabla)$.

Let (x^n) be a sequence in $h_M(\nabla)$ such that $\|x^n - x\| \to 0$, as $n \to \infty$, where $x \in \ell_M(\nabla)$.

There is some m corresponding to each $\rho > 0$ such that

$$\|x^m - x\| \leq \frac{\rho}{2}.$$

For each $\rho > 0$, using the convexity of M, we get

$$\sum_i M\left(\frac{|(\nabla x)_i|}{\rho}\right) = \sum_i M\left(\frac{2|(\nabla x)_i^m| - 2(|(\nabla x)_i^m| - |(\nabla x)_i|)}{2\rho}\right)$$

$$\leq \frac{1}{2}\sum_i M\left(\frac{2|(\nabla x)_i^m|}{\rho}\right) + \frac{1}{2}\sum_i M\left(\frac{2|(\nabla(x^m - x))_i|}{\rho}\right)$$

$$\leq \frac{1}{2}\sum_i M\left(\frac{2|(\nabla x)_i^m|}{\rho}\right) + \frac{1}{2}\sum_i M\left(\frac{2|(\nabla(x^m - x))_i|}{\|x^m - x\|}\right) < \infty$$

by Proposition 5.2.6. Hence, we have the following result:

Theorem 5.2.12. *Let M be an Orlicz function. Then, $(h_M(\nabla), \|\cdot\|)$ is an AK-BK-space.*

It is clear from the preceding result that $\ell_M(\nabla)$ is not an AK-space, in general. If an Orlicz function M satisfies the Δ_2 condition at 0, then the space $\ell_M(\nabla)$ reduces to $h_M(\nabla)$. Therefore, we have the following proposition:

Proposition 5.2.13. *If an Orlicz function M satisfies the Δ_2 condition at 0, then $\ell_M(\nabla)$ is an AK-space.*

Proof. It is enough to prove that $\ell_M(\nabla) \subseteq h_M(\nabla)$, when M satisfies the Δ_2-condition (see [64, p. 305]). □

It is also easy to establish the following two results using similar techniques available in Kamthan and Gupta [64] and Dutta [51]:

Proposition 5.2.14. *If M_1 and M_2 are two equivalent Orlicz functions, then $\ell_{M_1}(\nabla) = \ell_{M_2}(\nabla)$ and the following identity map is a topological isomorphism:*

$$I : \left(\ell_{M_1}(\nabla), \|\cdot\|_{M_1}^{\nabla}\right) \to \left(\ell_{M_2}(\nabla), \|\cdot\|_{M_2}^{\nabla}\right).$$

Proposition 5.2.15. *Let M be an Orlicz function and k be the corresponding kernel. If $k(x) = 0$ for all $x \in [0, a]$, where a is some positive number, then $\ell_M(\nabla)$ and $h_M(\nabla)$ are topologically isomorphic to $\ell_\infty(\nabla)$ and $c_0(\nabla)$, respectively, where*

$$\ell_\infty(\nabla) := \left\{x = (x_i) \in \omega : \sup_{i\in\mathbb{N}} |(\nabla x)_i| < \infty\right\}$$

and

$$c_0(\nabla) := \left\{ x = (x_i) \in \omega : \lim_{i\to\infty} (\nabla x)_i = 0 \right\}.$$

5.3 Orlicz Sequence Spaces Generated by Cesàro Mean

The results of this section are inspired by the findings of the previous section. In this section, we shall introduce a generalized version of Orlicz sequence space by using Cesàro mean of order one and study several structural properties by introducing different norms. The same approach can be adopted to construct several other generalized versions of Orlicz sequence spaces by using different other summability means. The proofs of the results can be obtained by using similar arguments as applied in the corresponding results of the previous section and for more details, we also refer to Dutta and Başar [52].

Let us consider the Cesàro matrix $C = (c_{nk})$ of order one, defined in Chapter 1. We begin with the following two definitions due to Dutta and Başar [52]:

Definition 5.3.1. *Let M be an Orlicz function and $\delta(M,x) = \sum_i M(|x_i|)$, where $x = (x_i) \in \omega$. Then, $\widetilde{\ell}_M(C)$ and $\widetilde{\ell}_M$ are defined as*

$$\widetilde{\ell}_M(C) := \left\{ x = (x_i) \in \omega : \widetilde{\delta}_C(M,x) = \sum_i M\left(\frac{1}{i+1} \left| \sum_{j=0}^{i} x_j \right| \right) < \infty \right\}$$

and

$$\widetilde{\ell}_M := \{ x \in \omega : \delta(M,x) < \infty \}.$$

Definition 5.3.2. *Let M and Φ be mutually complementary Orlicz functions. Then, the set $\ell_M(C)$ is defined as*

$$\ell_M(C) := \left\{ x = (x_i) \in \omega : \sum_i \left(\frac{1}{i+1} \sum_{j=0}^{i} x_j \right) y_i \text{ converges for all } y = (y_i) \in \widetilde{\ell}_\Phi \right\}.$$

Using Young's inequality, it is easy to see that $\widetilde{\ell}_M(C) \subseteq \ell_M(C)$ holds for any Orlicz function M. Also, the following quantity is finite

$$\sup \left\{ \left| \sum_i \left(\frac{1}{i+1} \sum_{j=0}^{i} x_j \right) y_i \right| : \delta(\Phi, y) \le 1 \right\}$$

for all $x \in \ell_M(C)$.

Suppose to the contrary that the above quantity is not finite. Then, for each $n \in \mathbb{N}$, there exists y^n with $\delta(\Phi, y^n) \leq 1$ such that

$$\left| \sum_i \left(\frac{1}{i+1} \sum_{j=0}^{i} x_j \right) y_i^n \right| > 2^{n+1}.$$

Assume without loss of generality that $x, y^n \geq 0$ for all $n \in \mathbb{N}$ and define $z = (z_i)$, as follows

$$z_i = \sum_n \frac{y_i^n}{2^{n+1}}$$

for all $i \in \mathbb{N}$. Let l be any positive integer. Using the convexity of Φ, we have

$$\begin{aligned} \Phi\left(\sum_{n=0}^{l} \frac{1}{2^{n+1}} y_i^n \right) &\leq \frac{1}{2} \left[\Phi(y_i^0) + \Phi\left(y_i^1 + \frac{y_i^2}{2} + \cdots + \frac{y_i^l}{2^{l-1}} \right) \right] \\ &\vdots \\ &\leq \sum_{n=0}^{l} \frac{1}{2^{n+1}} \Phi(y_i^n). \end{aligned}$$

By the continuity of Φ, we have

$$\delta(\Phi, z) = \sum_i \Phi(z_i) \leq \sum_i \sum_n \frac{1}{2^{n+1}} \Phi(y_i^n) \leq \sum_n \frac{1}{2^{n+1}} = 1.$$

But for every $l \in \mathbb{N}$, we get

$$\begin{aligned} \sum_i \left(\frac{1}{i+1} \sum_{j=0}^{i} x_j \right) z_i &\geq \sum_i \left(\frac{1}{i+1} \sum_{j=0}^{i} x_j \right) \sum_{n=0}^{l} \frac{1}{2^{n+1}} y_i^n \\ &= \sum_{n=0}^{l} \sum_i \left(\frac{1}{i+1} \sum_{j=0}^{i} x_j \right) \frac{y_i^n}{2^{n+1}} \geq l. \end{aligned}$$

This implies that $x \notin \ell_M(C)$.

It can be proved by using the similar technique as applied in the previous section (or see [52]) that $\ell_M(C)$ is a normed linear space under the norm $\|\cdot\|_M^C$ defined by

$$\|x\|_M^C = \sup \left\{ \left| \sum_i \left(\frac{1}{i+1} \sum_{j=0}^{i} x_j \right) y_i \right| : \delta(\Phi, y) \leq 1 \right\}.$$

In fact, we can state the following results given by Dutta and Başar [52]:

Theorem 5.3.3. *The space* $\left(\ell_M(C)\|\cdot\|_M^C\right)$ *is a Banach space.*

Corollary 5.3.4. *The space* $\left(\ell_M(C)\|\cdot\|_M^C\right)$ *is a BK-space.*

Let us consider the following quantity

$$\inf\left\{\rho>0:\sum_i M\left(\frac{1}{\rho(i+1)}\sum_{j=0}^{i}|x_j|\right)\le 1\right\}$$

for $x\in\ell_M(C)$.

Let $x=(x_i)$ and $y=(y_i)$ be any two elements of $\ell_M(C)$. We choose $\rho_1,\ \rho_2>0$ such that

$$\sum_i M\left(\frac{1}{\rho_1(i+1)}\sum_{j=0}^{i}|x_j|\right)\le 1$$

and

$$\sum_i M\left(\frac{1}{\rho_2(i+1)}\sum_{j=0}^{i}|x_j|\right)\le 1.$$

On considering $\rho=\rho_1+\rho_2$ and using the convexity of M yields

$$\begin{aligned}\sum_i M\left(\frac{1}{\rho(i+1)}\sum_{j=0}^{i}|x_j|\right) &\le \frac{\rho_1}{\rho_1+\rho_2}\sum_i M\left(\frac{1}{\rho_1(i+1)}\sum_{j=0}^{i}|x_j|\right)\\ &\quad+\frac{\rho_2}{\rho_1+\rho_2}\sum_i M\left(\frac{1}{\rho_2(i+1)}\sum_{j=0}^{i}|x_j|\right)\\ &\le 1.\end{aligned}$$

Hence,

$$\begin{aligned}&\inf\left\{\rho>0:\sum_i M\left(\frac{1}{\rho(i+1)}\sum_{j=0}^{i}|(x_j+y_j)|\right)\le 1\right\}\\ &\le \inf\left\{\rho_1>0:\sum_i M\left(\frac{1}{\rho_1(i+1)}\sum_{j=0}^{i}|x_j|\right)\le 1\right\}\\ &+\inf\left\{\rho_2>0:\sum_i M\left(\frac{1}{\rho_2(i+1)}\sum_{j=0}^{i}|x_j|\right)\le 1\right\}.\end{aligned}$$

Proposition 5.3.5. *The space* $\ell_M(C)$ *is a normed linear space under the norm* $\|\cdot\|_{(M)}^C$ *defined as*

$$\|x\|_{(M)}^C = \inf\left\{\rho > 0 : \sum_i M\left(\frac{1}{\rho(i+1)}\sum_{j=0}^{i}|x_j|\right) \leq 1\right\}.$$

Proof. The last inequality gives

$$\|x+y\|_{(M)}^C \leq \|x\|_{(M)}^C + \|y\|_{(M)}^C$$

for any $x, y \in \ell_M(C)$.

The other norm axioms can be obtained using similar arguments as applied for the norm $\|\cdot\|_{(M)}^{\nabla}$ of the previous section. □

We are now inspired to define the following space:

Definition 5.3.6. *Let* M *be any Orlicz function. Following* Dutta and Başar [52], *we define* $\ell_M'(C)$ *by*

$$\ell_M'(C) := \left\{x = (x_i) \in \omega : \sum_i M\left(\frac{1}{\rho(i+1)}\sum_{j=0}^{i}|x_j|\right) < \infty \text{ for some } \rho > 0\right\}.$$

Proposition 5.3.7. *The following statements hold:*

(i) $\left(\ell_M'(C), \|\cdot\|_{(M)}^C\right)$ *is a normed linear space.*

(ii) $\left(\ell_M'(C), \|\cdot\|_{(M)}^C\right)$ *is a Banach space.*

(iii) $\left(\ell_M'(C), \|\cdot\|_{(M)}^C\right)$ *is a BK-space.*

Proof. (i) The proof is similar as that for Proposition 5.3.5.

(ii) Let (x^s) be a Cauchy sequence in $\ell_M'(C)$. Let $\delta > 0$ be fixed and $q > 0$ be given such that $0 < \epsilon < 1$ and $q\delta \geq 1$. Then, there exists a positive integer n_0 such that $\|x^s - x^t\|_{(M)}^C < \frac{\epsilon}{q\delta}$ for all $s, t \geq n_0$. It follows that

$$\sum_i M\left(\frac{1}{\|x^s - x^t\|_{(M)}^C (i+1)}\sum_{j=0}^{i}\left|(x_j^s - x_j^t)\right|\right) \leq 1$$

for all $s, t \geq n_0$ and $i \in \mathbb{N}$.

Choose $q > 0$ with $M\left(\frac{q\delta}{2}\right) \geq 1$ so that

$$\sum_i M\left(\frac{1}{\|x^s - x^t\|_{(M)}^C (i+1)}\sum_{j=0}^{i}\left|(x_j^s - x_j^t)\right|\right) \leq M\left(\frac{q\delta}{2}\right)$$

for all $s, t \geq n_0$ and $i \in \mathbb{N}$.

Since M is non-decreasing, it follows that (x_i^s) is convergent in $\mathbb{C}$ for each $i \in \mathbb{N}$. Let $\lim_{s\to\infty} x_i^s = x_i$ for each $i \in \mathbb{N}$. It is now easy to see that $(x^s - x) \in \ell'_M(C)$ by using the continuity of Orlicz function and modulus. So, the linearity of $\ell'_M(C)$ gives that $x = (x_i) \in \ell'_M(C)$.

(iii) It is easy to conclude from the proof of Part (ii) that $\|x^s\|_{(M)}^C \to 0$ as $s \to \infty$ implies that $x_i^s \to 0$ as $i \to \infty$ for each $s \in \mathbb{N}$. □

The following result is immediate by the definition of the norm $\|\cdot\|_{(M)}^C$:

Proposition 5.3.8. *The following inequality*

$$\sum_i M\left(\frac{1}{\|x\|_{(M)}^C (i+1)} \sum_{j=0}^{i} |x_j|\right) \leq 1$$

holds for all $x = (x_i) \in \ell'_M(C)$.

Let M be an Orlicz function. We can define a subspace $h_M(C)$ of $\ell'_M(C)$, as follows:

$$h_M(C) := \left\{ x = (x_i) \in \omega : \sum_i M\left(\frac{1}{\rho(i+1)} \sum_{j=0}^{i} |x_j|\right) < \infty \text{ for each } \rho > 0 \right\}.$$

The topology of $h_M(C)$ is induced by $\|\cdot\|_{(M)}^C$. Henceforth, we shall write $\|\cdot\|$ instead of $\|\cdot\|_{(M)}^C$ provided it does not lead to any confusion.

Proposition 5.3.9. *The following statements hold:*

(i) $(h_M^C, \|\cdot\|)$ is an AK-BK-space, for any Orlicz function M.

(ii) If an Orlicz function M satisfies the Δ_2-condition at 0, then $\ell'_M(C)$ is an AK-space.

Proof. (i) Let $\epsilon \in (0,1)$. Then it can be shown that $\|x - x^{(n)}\| < \epsilon$ and also $h_M(C)$ is a closed subspace of $\ell'_M(C)$ by using similar technique as applied in the previous section.

(ii) It is enough to show that $h_M(C) \subseteq \ell'_M(C)$, if M satisfies the Δ_2-condition at 0 (see [52]). This will imply that $h_M(C) = \ell'_M(C)$, if M satisfies the Δ_2-condition at 0, and consequently $\ell'_M(C)$ is an AK-space. □

Proposition 5.3.10. *The following statements hold:*

(i) If M_1 and M_2 are equivalent Orlicz functions, then $\ell'_{M_1}(C) = \ell'_{M_2}(C)$ and the identity map

$$I : \left(\ell'_{M_1}(C), \|\cdot\|_{(M_1)}^C\right) \to \left(\ell'_{M_2}(C), \|\cdot\|_{(M_2)}^C\right)$$

is a topological isomorphism.

(ii) Let M be an Orlicz function and k be the corresponding kernel. If $k(x) = 0$ for all $x \in [0, a]$, where a is some positive number, then the spaces $\ell'_M(C)$ and $h_M(C)$ are topologically isomorphic to the spaces $\ell_\infty(C)$ and $c_0(C)$, respectively, where

$$\ell_\infty(C) := \left\{ x = (x_i) \in \omega : \sup_{i \in \mathbb{N}} \frac{1}{i+1} \sum_{j=0}^{i} |x_j| < \infty \right\}$$

and

$$c_0(C) := \left\{ x = (x_i) \in \omega : \lim_{i \to \infty} \frac{1}{i+1} \sum_{j=0}^{i} x_j = 0 \right\}.$$

Proof. (i) Let M_1 and M_2 be two equivalent Orlicz functions. Then, there exist positive constants α, β and b such that

$$M_1(\alpha u) \leq M_2(u) \leq M_1(\beta u) \text{ for all } u \in [0, b].$$

Let $x = (x_i) \in \ell'_{M_2}(C)$. Then,

$$\sum_i M_2 \left(\frac{1}{\rho(i+1)} \sum_{j=0}^{i} |x_j| \right) < \infty \text{ for some } \rho > 0.$$

For some $l > 0$ with $\frac{1}{l\rho(i+1)} \sum_{j=0}^{i} |x_j| \leq b$ for each $i \in \mathbb{N}$, we get

$$\sum_i M_1 \left(\frac{\alpha}{l\rho(i+1)} \sum_{j=0}^{i} |x_j| \right) \leq \sum_i M_2 \left(\frac{1}{\rho(i+1)} \sum_{j=0}^{i} |x_j| \right) < \infty.$$

So, the inclusion $\ell'_{M_2}(C) \subseteq \ell'_{M_1}(C)$ holds. In the same way, it can be seen that the inclusion $\ell'_{M_1}(C) \subseteq \ell'_{M_2}(C)$ also holds.

Next, we shall abbreviate $\|\cdot\|^C_{(M_1)}$ and $\|\cdot\|^C_{(M_2)}$ as $\|\cdot\|_1$ and $\|\cdot\|_2$, respectively.

We have

$$\sum_i M_2 \left(\frac{1}{\|x\|_2(i+1)} \sum_{j=0}^{i} |x_j| \right) \leq 1$$

for any $x = (x_i) \in \ell'_{M_2}(C)$.

It is possible to find $\mu > 1$ with $\left(\frac{b}{2}\right) \mu k_2 \left(\frac{b}{2}\right) \geq 1$, where k_2 is the kernel associated with M_2. Hence,

$$\sum_i M_2 \left(\frac{1}{\|x\|_2(i+1)} \sum_{j=0}^{i} |x_j| \right) \leq \left(\frac{b}{2}\right) \mu k_2 \left(\frac{b}{2}\right)$$

for all $i \in \mathbb{N}$. This implies that

$$\frac{1}{\mu\|x\|_2(i+1)}\sum_{j=0}^{i}|x_j| \leq b$$

for all $i \in \mathbb{N}$. Therefore,

$$\sum_i M_1\left(\frac{\alpha}{\mu\|x\|_2(i+1)}\sum_{j=0}^{i}|x_j|\right) \leq 1.$$

Thus, $\|x\|_1 \leq \frac{\mu}{\alpha}\|x\|_2$. It can be shown similarly that $\|x\|_2 \leq \beta\gamma\|x\|_1$ by choosing γ with $\gamma\beta > 1$ such that $\gamma\beta(\frac{b}{2})k_1(\frac{b}{2}) \geq 1$. Hence, $\frac{\alpha}{\mu}\|x\|_1 \leq \|x\|_2 \leq \beta\gamma\|x\|_1$, establishing that I is a topological isomorphism.

(ii) It is now easy to establish this part following similar arguments as applied in Kamthan and Gupta [64, p. 310]. □

5.4 Generalized Modular Sequence Spaces

Modular sequence spaces are generalizations of Orlicz sequence spaces. The concept introduced by Nakano [107] regarding modular sequence space was further generalized by Woo [132] by defining the space $\ell\{M_i\}$ as follows:

For a sequence of Orlicz functions $\{M_i\}$, the vector space $\ell\{M_i\}$ is defined by

$$\ell\{M_i\} := \left\{x = (x_i) \in \omega : \sum_{i=1}^{\infty} M_i\left(\frac{|x_i|}{\rho}\right) < \infty \text{ for some } \rho > 0\right\}.$$

The space $\ell\{M_i\}$ becomes a Banach space under the following norm

$$\|x\| = \inf\left\{\rho > 0 : \sum_{i=1}^{\infty} M_i\left(\frac{|x_i|}{\rho}\right) \leq 1\right\}$$

and is called a modular sequence space.

Earlier Nakano [107] introduced the space $\ell\{M_i\}$ for $M_i(x) = x^{\alpha_i}$, where $1 \leq \alpha_i < \infty$ for $i \in \mathbb{N}_1$. The results on generalized Orlicz sequence spaces as discussed in the previous two sections should be extendable to modular sequence spaces provided the generalized notions of equivalence and the Δ_2-condition at 0 for sequences of Orlicz functions are handled appropriately.

Dutta and Jebril [53] studied a more generalized class of modular sequence spaces by using Cesàro mean of order one and a sequence of positive real numbers and by considering a seminormed space as the base space as follows:

Let (M_i) be a sequence of Orlicz functions, let (X, q) be a seminormed space, let $p = (p_i)$ be a sequence of positive real numbers, and let C be the Cesàro matrix of order one. Then for nonnegative real numbers s, the set $\ell\{M_i, p, q, s, C\}$ is defined as follows:

$$\ell\{M_i, p, q, s, C\} :=$$
$$\left\{ x \in \omega(X) : \sum_i i^{-s} \left[M_i \left(q \left(\frac{1}{\rho(i+1)} \sum_{j=0}^{i} x_j \right) \right) \right]^{p_i} < \infty \text{ for some } \rho > 0 \right\}.$$

Let $p = (p_i)$ be a positive sequence of real numbers with $0 < p_i \leq \sup p_i = G$, $D = \max\{1, 2^{G-1}\}$ then the following inequality holds

$$|a_i + b_i|^{p_i} \leq D\{|a_i|^{p_i} + |b_i|^{p_i}\} \tag{5.1}$$

for all $a_i, b_i \in \mathbb{C}$ and $i \in \mathbb{N}$.

If we choose $p = (p_i)$ to be a bounded sequence of positive reals, then it is easy to see that $\alpha x + \beta y \in \ell\{M_i, p, q, s, C\}$, whenever $x, y \in \ell\{M_i, p, q, s, C\}$ and $\alpha, \beta \in \mathbb{C}$, by using inequality (5.1). Thus, $\ell\{M_i, p, q, s, C\}$ is a linear space over the field of complex numbers provided $p = (p_i)$ is a bounded sequence of positive reals.

Let us define the function g on $\ell\{M_i, p, q, s, C\}$ as follows ([53]):

$$g(x) = \inf \left\{ \rho^{\frac{p_n}{H}} : \sum_i i^{-s} \left[M_i \left(q \left(\frac{1}{\rho(i+1)} \sum_{j=0}^{i} x_j \right) \right) \right]^{p_i} \leq 1 \text{ for all } n \in \mathbb{N}_1 \right\},$$

where $H = \max\{1, \sup_i p_i\}$.

Then using similar arguments as applied in Proposition 5.3.5, it can be shown for $x, y \in \ell\{M_i, p, q, s, C\}$ and a non-zero scalar λ that

$$g(x+y) \leq g(x) + g(y) \text{ and } g(\lambda x) = |\lambda| g(x).$$

Also, it is obvious that $g(x) = g(-x)$ and $g(\theta) = 0$.

Thus, we have the following result:

Proposition 5.4.1. *Let $p = (p_i)$ be a bounded sequence of positive reals. Then $(\ell\{M_i, p, q, s, C\}, g)$ is a paranormed space (need not be total paranorm).*

The construction of the space $\ell\{M_i, p, q, s, C\}$ indicates that it may have different copies several other spaces can be derived from this space under different conditions. Next, we discuss some results on these aspects.

Let (M_i) and (T_i) be two sequences of Orlicz functions, let $p = (p_i)$ and $t = (t_i)$ be two bounded sequences of positive reals, and let q_1 and q_1 be two seminormed on the same space X. Then it is easy to see that ([53])

(i) $\ell\{M_i, p, q_1, s, C\} \bigcap \ell\{M_i, p, q_2, s, C\} \neq \emptyset$,

(ii) $\ell\{M_i, p, q_1, s, C\} \bigcap \ell\{M_i, p, q_2, s, C\} \subseteq \ell\{M_i, p, q_1 + q_2, s, C\}$,

(iii) $\ell\{M_i, p, q, s, C\} \bigcap \ell\{T_i, p, q, s, C\} \subseteq \ell\{M_i + T_i, p, q, s, C\}$,

(iv) $\ell\{M_i, p, q_1, s, C\} \subseteq \ell\{M_i, p, q_2, s, C\}$, whenever q_1 is stronger than q_2,

(v) $\ell\{M_i, p, q, s_1, C\} \subseteq \ell\{M_i, p, q, s_2, C\}$, whenever $s_1 \leq s_2$.

If we take $s = 0$, then the space $\ell\{M_i, p, q, s, C\}$ reduces to the space $\ell\{M_i, p, q, C\}$, where

$$\ell\{M_i, p, q, C\} := \left\{ x \in \omega(X) : \sum_i \left[M_i \left(q \left(\frac{1}{\rho(i+1)} \sum_{j=0}^{i} x_j \right) \right) \right]^{p_i} < \infty \text{ for some } \rho > 0 \right\}.$$

Using similar technique of Part (ii) of Proposition 5.3.7, the following result can be established (see also, [53]):

Proposition 5.4.2. *Let $p = (p_i)$ be a bounded sequence of positive reals, and let (X, q) be a complete seminormed space. Then $(\ell\{M_i, p, q, C\}, h)$ is a complete paranormed space, where h is defined by*

$$h(x) = \inf \left\{ \rho^{\frac{p_n}{H}} : \sum_i \left[M_i \left(q \left(\frac{1}{\rho(i+1)} \sum_{j=0}^{i} x_j \right) \right) \right]^{p_i} \leq 1 \textit{ for all } n \in \mathbb{N}_1 \right\},$$

where $H = \max\{1, \sup_i p_i\}$.

If we take $s = 0$ and $p_i = a$, a constant for all $i \in \mathbb{N}$, then the space $\ell\{M_i, p, q, s, C\}$ reduces to the space $\ell\{M_i, q, C\}$, where

$$\ell\{M_i, q, C\} := \left\{ x \in \omega(X) : \sum_i \left[M_i \left(q \left(\frac{1}{\rho(i+1)} \sum_{j=0}^{i} x_j \right) \right) \right] < \infty \text{ for some } \rho > 0 \right\}.$$

Let us define the function $\|\cdot\|_{\ell\{M_i,q,C\}}$ on $\ell\{M_i, q, C\}$, as follows (see [53]):

$$\|x\|_{\ell\{M_i,q,C\}} = \inf \left\{ \rho : \sum_i \left[M_i \left(q \left(\frac{1}{\rho(i+1)} \sum_{j=0}^{i} x_j \right) \right) \right] \leq 1 \right\}.$$

We assume that (X, q) is a normed space for the next result.

Proposition 5.4.3. *We have the following:*

(i) $(\ell\{M_i, q, C\}, \|\cdot\|_{\ell\{M_i,q,C\}})$ is a normed space.

(ii) If (X, q) is a complete normed space, then $(\ell\{M_i, q, C\}, \|\cdot\|_{\ell\{M_i,q,C\}})$ is a Banach space.

Proof. (i) The proof can be obtained using similar arguments as that of Theorem 5.2.5.

(ii) The proof can be obtained using similar arguments as that of Proposition 5.3.7 (ii). □

5.5 Conclusion

In this chapter, we discussed some possible ways to construct generalized Orlicz sequence spaces and studied their relevant algebraic and topological properties. Similar spaces can be constructed by considering higher order difference operators as well as in combination with other summability means. The primary aim of this chapter is to demonstrate a systematic approach to develop theory for generalized Orlicz sequence spaces, and the results presented are not comprehensive. Still there are many scopes to study the spaces introduced in this chapter. Lindenstrauss and Tzafriri ([79, 80]) pointed out a possible generalization of the Orlicz sequence space ℓ_M to the case when M is an Orlicz function that does not satisfy the convexity condition. An Orlicz function which does not satisfy the convexity condition is called a K-function. Kalton [63] picked up the problem and succeeded in finding many interesting features distinguishing these two theories of sequences spaces. Dutta and Kočinac [54] discussed a more general class of difference sequences by considering a sequence of K-functions. Dutta [50] computed algebraic duals for a class of difference sequences defined by Orlicz functions. There are still ample scopes to construct more generalized modular sequence spaces and sequence spaces by K-functions in combination with different summability methods and study their algebraic and topological properties, and also compute their continuous and algebraic duals. It is still to develop a rich theory to characterize matrix transformations involving Orlicz sequence spaces. The results discussed in this chapter may also be taken up for study for double sequences, and it is possible to develop a system approach to deal with Orlicz and modular spaces for double sequences, and also in combination with different summability techniques.

Bibliography

[1] Z.U. Ahmad, Mursaleen, *Köthe-Toeplitz duals of some new sequence spaces and their matrix maps*, Publ. Inst. Math. (Beograd) **42** (1987), 57–61.

[2] A.M. Akhmedov, S.R. El-Shabrawy, *On the fine spectrum of the operator $\Delta_{a,b}$ over the sequence space c*, Comput. Math. Appl. **61** (10) (2011), 2994–3002.

[3] B. Altay, *On the space of p-summable difference sequences of order m*, $(1 \leq p < \infty)$, Stud. Sci. Math. Hungar. **43** (4) (2006), 387–402.

[4] B. Altay, F. Başar, *On the paranormed Riesz sequence spaces of non-absolute type*, Southeast Asian Bull. Math. **26** (5) (2002), 701–715.

[5] B. Altay, F. Başar, *Some Euler sequence spaces of non-absolute type*, Ukrainian Math. J. **57** (1) (2005), 1–17.

[6] B. Altay, F. Başar, *Some paranormed Riesz sequence spaces of non-absolute type*, Southeast Asian Bull. Math. **30** (5) (2006), 591–608.

[7] B. Altay, F. Başar, *Some paranormed sequence spaces of non-absolute type derived by weighted mean*, J. Math. Anal. Appl. **319** (2) (2006), 494–508.

[8] B. Altay, F. Başar, *Generalisation of the sequence space $\ell(p)$ derived by weighted mean*, ibid., **330** (1) (2007), 174–185.

[9] B. Altay, F. Başar, *Certain topological properties and duals of the matrix domain of a triangle matrix in a sequence space*, ibid., **336** (1) (2007), 632–645.

[10] B. Altay, F. Başar, *The matrix domain and the fine spectrum of the difference operator Δ on the sequence space ℓ_p*, $(0 < p < 1)$, Commun. Math. Anal. **2** (2) (2007), 1–11.

[11] B. Altay, F. Başar, M. Mursaleen, *On the Euler sequence spaces which include the spaces ℓ_p and ℓ_∞ I*, Inform. Sci. **176** (10) (2006), 1450–1462.

[12] C. Aydın, B. Altay, *Domain of generalized difference matrix $B(r,s)$ on some Maddox's spaces*, Thai J. Math. **11** (1) (2013), 87–102.

[13] C. Aydın, F. Başar, *On the new sequence spaces which include the spaces* c_0 *and* c, Hokkaido Math. J. **33** (2) (2004), 383–398.

[14] C. Aydın, F. Başar, *Some new paranormed sequence spaces*, Inform. Sci. **160** (1-4) (2004), 27–40.

[15] C. Aydın, F. Başar, *Some new difference sequence spaces*, Appl. Math. Comput. **157** (3) (2004), 677–693.

[16] C. Aydın, F. Başar, *Some new sequence spaces which include the spaces* ℓ_p *and* ℓ_∞, Demonstratio Math. **38** (3) (2005), 641–656.

[17] C. Aydın, F. Başar, *Some generalizations of the sequence space* a_p^r, Iran. J. Sci. Technol. Trans. A, Sci. **30** (2006), No. A2, 175–190.

[18] C. Aydın, F. Başar, *Some topological and geometric properties of the domain of the generalized difference matrix* $B(r,s)$ *in the sequence space* $\ell(p)$, Thai J. Math. **12** (1) (2014), 113–132.

[19] S. Banach, *Theorie des Operations Lineaires*, Warzawa, 1932.

[20] F. Başar, *Infinite matrices and almost boundedness*, Boll. Unione Mat. Ital. A (7) **6** (1992), 395–402.

[21] F. Başar, *Summability Theory and Its Applications*, Bentham Science Publishers, e-books, Monographs, İstanbul, 2012.

[22] F. Başar, *Domain of the composition of some triangles in the space of p-summable sequences*, AIP Conference Proceedings, **1611** (2014), 348–356.

[23] F. Başar, *Domain of the composition of some triangles in the space of p-summable sequences for* $0 < p \leq 1$, AIP Conference Proceedings **1759** (2016), 020003-1-020003-6; doi: 10.1063/1.4959617.

[24] F. Başar, B. Altay, *Matrix mappings on the space* $bs(p)$ *and its* $\alpha-$, $\beta-$ *and* $\gamma-$*duals*, Aligarh Bull. Math. **21** (1) (2002), 79–91.

[25] F. Başar, B. Altay, *On the space of sequences of p-bounded variation and related matrix mappings*, (English, Ukrainian summary) Ukrain. Mat. Zh. **55** (1) (2003), 108–118; reprinted in Ukrainian Math. J. **55** (1) (2003), 136–147.

[26] F. Başar, B. Altay, M. Mursaleen, *Some generalizations of the space* bv_p *of p-bounded variation sequences*, Nonlinear Anal. **68** (2) (2008), 273–287.

[27] F. Başar, N.L. Braha, *Euler-Cesàro difference spaces of bounded, convergent and null sequences*, Tamkang J. Math. **47** (4) (2016), 405–420.

[28] F. Başar, A.F. Çakmak, *Domain of triple band matrix* $B(r,s,t)$ *on some Maddox's spaces*, Ann. Funct. Anal. **3** (1) (2012), 32–48.

[29] F. Başar, M. Kirişçi, *Almost convergence and generalized difference matrix*, Comput. Math. Appl. **61** (3) (2011), 602–611.

[30] M. Başarır, *On some new sequence spaces and related matrix transformations*, Indian J. Pure Appl. Math. **26** (10) (1995), 1003–1010.

[31] M. Başarır, *On the generalized Riesz B-difference sequence spaces*, Filomat **24** (4) (2010), 35–52.

[32] M. Başarır, F. Başar, E.E. Kara, *On the spaces of Fibonacci difference null and convergent sequences*, Sarajevo J. Math. **12** (25) (2016), 167–182.

[33] M. Başarır, M. Kayıkçı, *On the generalized B^m-Riesz sequence space and β-property*, J. Inequal. Appl. 2009 (2009), Article ID 385029, 18 pp.

[34] M. Başarır, M. Öztürk, *On the Riesz difference sequence space*, Rend. Circ. Mat. Palermo (2) 57 (2008), no. 3, 377–389.

[35] J. Boos, *Classical and Modern Methods in Summability*, Oxford University Press Inc., New York, 2000.

[36] N.L. Braha, F. Başar, *On the domain of the triangle $A(\lambda)$ on the spaces of null, convergent and bounded sequences*, Abstr. Appl. Anal. **2013**, Article ID 476363, 9 pages, 2013, doi:10.1155/2013/476363.

[37] M. Candan, *Domain of the double sequential band matrix in the classical sequence spaces*, J. Inequal. Appl. (2012), 2012:281, 15p.

[38] M. Candan, *Almost convergence and double sequential band matrix*, Acta Math. Sci., Ser. B, Engl. Ed. **34** (2) (2014), 354–366.

[39] S. Chen, *Geometry of Orlicz spaces*, Diss. Math. **356** (1996), 1–224.

[40] B. Choudhary, S.K. Mishra, *On Köthe-Toeplitz duals of certain sequence spaces and thair matrix transformations*, Indian J. Pure Appl. Math. **24** (59) (1993), 291–301.

[41] R.G. Cooke, *Infinite Matrices and Sequence Spaces*, Macmillan and Co. Limited, London, 1950.

[42] H. Coşkun, F. Başar, *On the characterization of certain matrix classes*, Boll. Unione Mat. Ital. A(7) **8** (2) (1994), 215–222.

[43] A.F. Çakmak, F. Başar, *Some sequence spaces and matrix transformations in multiplicative sense*, TWMS J. Pure Appl. Math. **6** (1) (2015), 27–37.

[44] R. Çolak, Ö. Çakar, *Banach limits and related matrix transformations*, Stud. Sci. Math. Hung. **24** (1989), 429–436.

[45] R. Çolak, M. Et, *On some generalized difference sequence spaces and related matrix transformations*, Hokkaido Math. J. **26** (3) (1997), 483–492.

[46] G. Das, *Banach and other limits*, J. London Math. Soc. **7** (1973), 501–507.

[47] J. Diestel, *Geometry of Banach Spaces–Selected Topics*, Springer, Berlin, Germany, 1984.

[48] J.P. Duran, *Infinite matrices and almost convergence*, Math. Z. **128** (1972), 75–83.

[49] M. Et, R. Çolak, *On some generalized difference sequence spaces*, Soochow J. Math. **21** (4) (1995), 377–386.

[50] H. Dutta, *On Köthe-Toeplitz and null duals of some difference sequence spaces defined by Orlicz functions*, Eur. J. Pure Appl. Math. **2** (4) (2009), 554–463.

[51] H. Dutta, *On Orlicz difference sequence spaces*, SDU J. Sci. **5** (1) (2010), 119–136.

[52] H. Dutta, F. Başar, *A generalization of Orlicz sequence spaces by Cesàro mean of order one*, Acta Math. Univ. Comen. **80** (2) (2011), 185–200.

[53] H. Dutta, I.H. Jebril, *An extension of modular sequence spaces*, Abstr. Appl. Anal. Art. Id. 371806 (2013), 1–7.

[54] H. Dutta, L.D.R. Kočinac, *On difference sequence spaces defined by Orlicz functions without convexity*, Bull. Iranian Math. Soc. **41** (2) (2014), 477–489.

[55] C. Eizen, G. Laush, *Infinite matrices and almost convergence*, Math. Japon. **14** (1969), 137–143.

[56] M. Et, *On some difference sequence spaces*, Turkish J. Math. **17** (1993), 18–24.

[57] Y. Gribanov, *On the theory of ℓ_M−spaces (Russian)*, Uc. Zap. Kazansk un-ta 117 (1957), 62–65.

[58] K. -G. Grosse-Erdmann, *Matrix transformations between the sequence spaces of Maddox*, J. Math. Anal. Appl. **180** (1993), 223–238.

[59] K.-G. Grosse-Erdmann, *On ℓ^1-invariant sequence spaces*, J. Math. Anal. Appl. **262** (2001), 112–132.

[60] A.M. Jarrah, E. Malkowsky, *BK spaces, bases and linear operators*, Rendiconti Circ. Mat. Palermo II **52** (1) (1998), 177–191.

[61] A.M. Jarrah, E. Malkowsky, *Ordinary, absolute and strong summability and matrix transformations*, Filomat **17** (2003), 59–78.

[62] M.A. Krasnoselskii, Y.B. Rutitsky, *Convex Functions and Orlicz Spaces*, Groningen, Netherlands, 1961.

[63] N.J. Kalton, *Orlicz sequence spaces without local convexity*, Proc. Cambridge Phil. Soc. **81** (2) (1977), 253–277.

[64] P.K. Kamthan, M. Gupta, *Sequence Spaces and Series*, Marcel Dekker Inc., New York and Basel, 1981.

[65] E.E. Kara, *Some topological and geometrical properties of new Banach sequence spaces*, J. Inequal. Appl. **2013:38**, 15 pages, 2013. doi:10.1186/1029-242X-2013-38.

[66] E.E. Kara, M. Başarır, M. Mursaleen, *Compact operators on the Fibonacci difference sequence spaces $\ell_p(\widehat{F})$ and $\ell_\infty(\widehat{F})$*, 1st International Eurasian Conference on Mathematical Sciences and Applications, Prishtine-Kosovo, September 3–7, 2012.

[67] E.E. Kara, M. Başarır, *On compact operators and some Euler $B(m)$-difference sequence spaces*, J. Math. Anal. Appl. **379** (2011), 499–511.

[68] E.E. Kara, M. Öztürk, M. Başarır, *Some topological and geometric properties of generalized Euler sequence spaces*, Math. Slovaca **60** (3) (2010), 385–398.

[69] K. Kayaduman, M. Şengönül, *The spaces of Cesàro almost convergent sequences and core theorems*, Acta Math. Sci. Ser. B Engl. Ed. **32** (6) (2012), 2265–2278.

[70] H. Kızmaz, *On certain sequence spaces*, Canad. Math. Bull. **24** (2) (1981), 169–176.

[71] J.P. King, *Almost summable sequences*, Proc. Amer. Math. Soc. **17** (1966), 1219–1225.

[72] M. Kirişci, *On the space of Euler almost null and Euler almost convergent sequences*, Commun. Fac. Sci. Univ. Ankara, Ser. A_1 **62** (2) (2013), 85–100.

[73] M. Kirişçi, F. Başar, *Some new sequence spaces derived by the domain of generalized difference matrix*, Comput. Math. Appl. **60** (5) (2010), 1299–1309.

[74] T. Koshy, *Fibonacci and Lucas Numbers with Applications*, Wiley, 2001.

[75] K. Lindberg, *Contractive Projections on Orlicz Sequences Spaces and Continuous Function Spaces*, Ph.D thesis, University of California, Berkeley, 1970

[76] K. Lindberg, *On subspaces of Orlicz sequence spaces*, Studia Math. **45** (1973), 119–146.

[77] C.G. Lascarides, I.J. Maddox, *Matrix transformations between some classes of sequences*, Proc. Camb. Phil. Soc. **68** (1970), 99–104.

[78] J. Lindenstrauss, L. Tzafriri, *On Orlicz sequence spaces*, Israel J. Math. **10** (3) (1971), 379–390.

[79] J. Lindenstrauss, L. Tzafriri, *On Orlicz sequence spaces II*, Israel J. Math. **11** (4) (1972), 355–379.

[80] J. Lindenstrauss, L. Tzafriri, *On Orlicz sequence spaces III*, Israel J. Math. **14** (4) (1973), 368–389.

[81] G.G. Lorentz, *A contribution to the theory of divergent sequences*, Acta Math. **80** (1948), 167–190.

[82] J. Musielak, *Orlicz Spaces and Modular Spaces*, Lecture Notes in Mathematics 1083, Springer-Verlag, 1983.

[83] I.J. Maddox, *On theorems of Steinhaus type*, J. London Math. Soc. **42** (1967), 239–244.

[84] I.J. Maddox, *Spaces of strongly summable sequences*, Quart. J. Math. Oxford (2), **18** (1967), 345–355.

[85] I.J. Maddox, *Paranormed sequence spaces generated by infinite matrices*, Proc. Camb. Phil. Soc. **64** (1968), 335–340.

[86] I.J. Maddox, *Some properties of paranormed sequence spaces*, London J. Math. Soc. (2) **1** (1969), 316–322.

[87] I.J. Maddox, *Some analogues of Knopp's core theorem*, Internat. J. Math. Math. Sci. **2** (1979), 605–614.

[88] I.J. Maddox, *Elements of Functional Analysis*, The University Press, Cambridge, second edition, 1988.

[89] I.J. Maddox, J.W. Roles, *Absolute convexity in certain topological linear spaces*, Proc. Cambridge Phil. Soc. **66** (1969), 541–545.

[90] L. Maligranda, *Orlicz Spaces and Interpolation*, Inst. Math. Polish Academy of Sciences, Poznan, 1985.

[91] E. Malkowsky, *Sequence Spaces, Matrix Transformations and Compact Operators*, unpublished Lecture Notes.

[92] E. Malkowsky, *Recent results in the theory of matrix transformations in sequence spaces*, Mat. Vesnik **49** (1997), 187–196.

[93] E. Malkowsky, *Linear operators between some matrix domains*, Rend. Circ. Mat. Palermo, II. Ser. **68** (2002), 641–655.

[94] E. Malkowsky, F. Başar, *A survey on some paranormed sequence spaces*, Filomat **31** (4) (2017), 1099–1122.

[95] E. Malkowsky, M. Mursaleen, S. Suantai, *The dual spaces of sets of difference sequences of order m and matrix transformations*, Acta Math. Sin. Eng. Ser. **23** (3) (2007), 521–532.

[96] E. Malkowsky, S.D. Parashar, *Matrix transformations in space of bounded and convergent difference sequences of order m*, Analysis **17** (1997), 87–97.

[97] E. Malkowsky, V. Rakočević, *An introduction into the theory of sequence spaces and measures of noncompactness*, Zbornik Radova, Matematički Institut SANU, Belgrade, **9 (17)** (2000), 143–234.

[98] E. Malkowsky, V. Rakočević, *On matrix domains of triangle*, Appl. Math. Comput. **189** (2) (2007), 1146–1163.

[99] E. Malkowsky, E. Savaş, *Matrix transformations between sequence spaces of generalized weighted means*, Appl. Math. Comput. **147** (2004), 333–345.

[100] F.M. Mears, *Some multiplication theorems for the Nörlund mean*, Bull. Amer. Math. Soc. **41** (12) (1935), 875–880.

[101] F.M. Mears, *The inverse Nörlund mean*, Ann. Math. **44** (3) (1943), 401–409.

[102] H.I. Miller, C. Orhan, *On almost convergent and statistically convergent subsequences*, Acta Math. Hung. **93** (2001), 135–151.

[103] Mursaleen, *Generalized spaces of difference sequences*, J. Math. Anal. Appl. **203** (3) (1996), 738–745.

[104] M. Mursaleen, *Elements of Metric Spaces*, Anamaya Publ. New Delhi, 2005.

[105] M. Mursaleen, A.K. Noman, *On the spaces of λ-convergent and bounded sequences*, Thai J. Math. **8** (2) (2010), 311–329.

[106] H. Nakano, *Concave modulars*, J. Math. Soc. Japan **5** (1953), 29–49.

[107] H. Nakano, *Modulared sequence spaces*, Proc. Japan Acad. **27** (2) (1951), 508–512.

[108] S. Nanda, *Matrix transformations and almost convergence*, Mat. Vesnik **13 (28)** (1976), no. 3, 305–312.

[109] S. Nanda, *Infinite matrices and almost convergence*, J. Indian Math. Soc. **40** (1976), 173–184.

[110] S. Nanda, *Matrix transformations and almost boundedness*, Glas. Mat. 14 (34) (1979), 99–107.

[111] S. Nanda, *Matrix Transformations and Sequence Spaces*, Internal report, I.C.T.P. (1983), IC/83/55.

[112] H. Nergiz, F. Başar, *Domain of the double sequential band matrix $B(\widetilde{r},\widetilde{s})$ in the sequence space $\ell(p)$*, Abstr. Appl. Anal. **2013**, Article ID 949282, 10 pages, 2013. doi: 10.1155/2013/949282.

[113] H. Nergiz, F. Başar, *Some topological and geometric properties of the domain of the double sequential band matrix $B(\widetilde{r},\widetilde{s})$ in the sequence space $\ell(p)$*, AIP Conference Proceedings **1470** (2012), 163–168.

[114] H. Nergiz, F. Başar, *Some geometric properties of the domain of the double sequential band matrix $B(\tilde{r},\tilde{s})$ in the sequence space $\ell(p)$*, Abstr. Appl. Anal. **2013**, Article ID 421031, 8 pages, 2013. doi:10.1155/2013/421031.

[115] P. -N. Ng, P. -Y. Lee, *Cesàro sequences spaces of non-absolute type*, Comment. Math. Prace Mat. **20** (2) (1978), 429–433.

[116] F. Özger, F. Başar, *Domain of the double sequential band matrix $B(\widetilde{r},\widetilde{s})$ on some Maddox's spaces*, AIP Conference Proceedings **1470** (2012), 152–155.

[117] F. Özger, F. Başar, *Domain of the double sequential band matrix $B(\widetilde{r},\widetilde{s})$ on some Maddox's spaces*, Acta Math. Sci., Ser. B, Engl. Ed. **34** (2) (2014), 394–408.

[118] B.L. Panigrahi, P.D. Srivastava, *Spectrum and fine spectrum of generalized second order difference operator Δ^2_{uv} on sequence space c_0*, Thai J. Math. **9** (1) (2011), 57–74.

[119] A. Peyerimhoff, *Lectures on Summability*, Lecture Notes in Mathematics, Springer-Verlag, Berlin, 1969.

[120] H. Polat, F. Başar, *Some Euler spaces of difference sequences of order m*, Acta Math. Sci. **27B** (2) (2007), 254–266.

[121] J.A. Sıddıqi, *Infinite matrices summing every almost periodic sequences*, Pac. J. Math. **39** (1) (1971), 235–251.

[122] S. Simons, *The sequence spaces $\ell(p_v)$ and $m(p_v)$*, Proc. London Math. Soc. **15** (3) (1965), 422–436.

[123] P.D. Srivastava, S. Kumar, *Fine spectrum of the generalized difference operator Δ_ν on sequence space ℓ_1*, Thai J. Math. **8** (2) (2010), 7–19.

[124] P.D. Srivastava, S. Kumar, *Fine spectrum of the generalized difference operator Δ_{uv} on sequence space l_1*, Appl. Math. Comput. **218** (11) (2012), 6407–6414.

[125] A. Sönmez, *Some new sequence spaces derived by the domain of the triple band matrix*, Comput. Math. Appl. **62** (2) (2011), 641–650.

[126] A. Sönmez, *Almost convergence and triple band matrix*, Math Comput. Model. **57** (9-10) (2013), 2393–2402.

[127] M. Stieglitz, H. Tietz, *Matrix transformationen von folgenräumen eine ergebnisübersicht*, Math. Z. **154** (1977), 1–16.

[128] M. Şengönül, F. Başar, *Some new Cesàro sequence spaces of non-absolute type which include the spaces c_0 and c*, Soochow J. Math. **31** (1) (2005), 107-119.

[129] M. Şengönül, K. Kayaduman, *On the Riesz almost convergent sequence space*, Abstr. Appl. Anal. **2012**, Art. ID 691694, 18 pp.

[130] O. Tuğ, F. Başar, *On the domain of Nörlund mean in the spaces of null and convergent sequences*, TWMS J. Pure Appl. Math. **7** (1) (2016), 76–87.

[131] O. Tuğ, F. Başar, *On the spaces of Nörlund almost null and Nörlund almost convergent sequences*, Filomat **30** (3) (2016), 773–783.

[132] J.Y.T. Woo, *On modular sequence spaces*, Studia Math. **48** (1973), 271–289.

[133] C.-S. Wang, *On Nörlund sequence space*, Tamkang J. Math. **9** (1978), 269–274.

[134] A. Wilansky, *Functional Analysis*, Blaisdell Publishing Company, New York, 1964.

[135] A. Wilansky, *Summability through Functional Analysis*, North-Holland Mathematics Studies **85**, Amsterdam, 1984.

[136] M. Yeşilkayagil, F. Başar, *Composite dual summability methods of the new sort*, Commun. Fac. Sci. Univ. Ankara, Ser. A_1 **62** (1) (2013), 33–43.

[137] M. Yeşilkayagil, F. Başar, *On the paranormed Nörlund sequence space of non-absolute type*, Abstr. Appl. Anal. **2014**, Article ID 858704, 9 pages, 2014. doi:10.1155/2014/858704.

[138] M. Yeşilkayagil, F. Başar, *Domain of the Nörlund matrix on some Maddox's spaces*, Proc. Natl. Acad. Sci. India, Sect. A Phys. Sci. **87** (2017), no. 3, 363–371. doi:10.1007/s40010-017-0359-4.

Index